MW00508485

THE OCTAVO

NATURE-PRINTED BRITISH FERNS.

VOL. I.—POLYPODIUM TO LASTREA.

GENUS I.—V.

LONDON:
BRADBURY & EVANS. 11, BOUVERIE ST
1859.

THE

OCTAVO NATURE-PRINTED

BRITISH FERNS:

BEING

FIGURES AND DESCRIPTIONS OF THE SPECIES AND VARIETIES
OF FERNS FOUND IN THE UNITED KINGDOM.

BY

THOMAS MOORE, F.L.S., F.H.S.,

Author of "The Ferns of Great Britain and Ireland Nature-Printed;" "The Handbook of British Ferns;" "Index Filicum," &c.; Curator of the Chelsea Botanic Garden.

NATURE-PRINTED BY HENRY BRADBURY.

IN TWO VOLUMES.

VOL. I.—POLYPODIUM TO LASTREA.

GENUS I.—V.

LONDON:
BRADBURY AND EVANS, 11, BOUVERIE STREET.
1859.

BRADBURY AND EVANS,
PRINTERS EXTRAORDINARY TO THE QUEEN,
WHITEFRIARS.

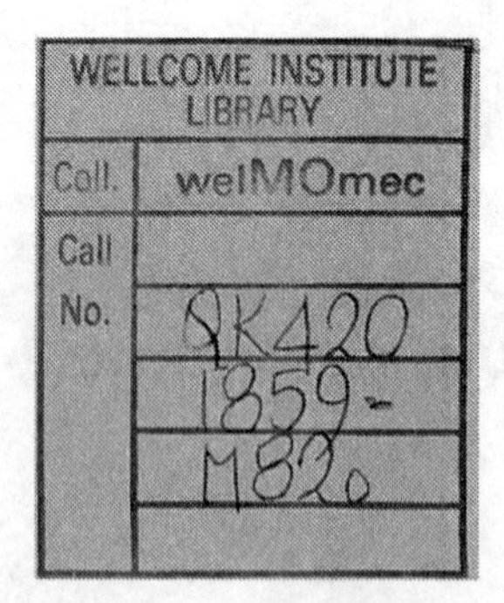

PREFACE.

In the Preface to the folio edition of this work, it was remarked that "everyone who has attempted to ascertain the name of a plant by comparing it with mere descriptions, is aware of the difficulty of effecting the object, unless he is more familiar with the technical language of science than those commonly are who, although admirers of delicate texture, beautiful colour, or graceful form, do not profess to be skilful Botanists. Even with the aid of drawings, investigation often leads to no satisfactory result, in consequence of the inability of art to represent faithfully the minute peculiarities by which natural objects are often best distinguished. If this is so with plants in general, it is most especially true of Ferns, the complicated forms and tender organisation of which baffle the most skilful and patient artist, who can only give at the best an imperfect sketch of what he supposes to be their more important features. And herein lies the great defect of all pictorial representations. The draughtsman can do no more than delineate a part of what he sees; and whether he sees correctly what he delineates will at all times be a matter of doubt, especially where, as in natural history, minute accuracy is indispensable. But if minute accuracy is of more vital importance in one than another race of plants, it is most especially so among Ferns, in the distinctions of which the form of indentations, general outline, the exact manner in which repeated subdivision is effected, and most especially the distribution of veins scarcely visible to the naked eye, play the most important part. To express such facts with the necessary accuracy, the art of a Talbot or a Daguerre was insufficient, nor could they be represented pictorially until NATURE-PRINTING was brought to its present state of perfection."

This art of Nature-Printing, which practice is gradually bringing nearer and nearer to a perfect state, represents not only the general form of such plants as Ferns, and others, with absolute accuracy, but also the veins, and the nature of the surface,—the hairs, and other minutiæ of superficial structure by which they are known, irrespective of the details of their fructification.

It is true that Nature-Printing has its defects as well as its advantages, for it, like the artist, can only represent a portion of the whole structure of the plant; but then its accuracy is perfect as far as it goes, and in the case of Ferns, it shows just that which it is most desirable to represent for practical purposes, that is, the outline, and the venation. If it fails, as it does, to give the details of the sori, and their indusia, it accurately gives the general form and arrangement even of those parts; and it must not be overlooked that such organs are not practically employed in the popular first-sight recognition of a Fern, although they are necessary subjects of examination in a scientific inquiry. The practised eye recognises at a glance the name of a Fern, not by looking to the form of its indusium, and the place occupied by the sori, but by its general manner of growth, its ramifications, and the form of its leaflets, all which peculiarities Nature-Printing shows with unerring truth. And this gives it its popular value: it sets forth correctly the first-sight appearance a plant bears, and thus, by familiarising the eye with its external features, enables even the tyro to recognise the prototype when it comes before him. It may indeed be assumed that in the Ferns, and many other plants, a knowledge of the inconspicuous parts of fructification might often be dispensed with, if it were possible accurately to represent by figures, or to describe by words, the real form and condition of the larger organs. On these points, good Nature-Printing conveys to the eye, positive and accurate impressions.

The present Work has been prepared, with the view of showing by unmistakeable evidence the differences which really exist among the Ferns which grow wild in Great Britain and Ireland. These beautiful plants have of late years attracted so much attention, and are now so universally cultivated, that it has become most desirable to establish upon solid grounds the true value of their characteristic marks—a result which it is hopeless to expect from mere descriptions or imperfect engravings.

This is not the place to discuss the soundness of the principles upon which the modern genera of Ferns are based; but we quite agree with the remark of Dr. Lindley, in the preface to the folio edition, that "the distribution of veins, and the position of sori with respect to them, are characters of equal importance with the form, or absence or presence of an indusium, or the direction in which it separates from the epidermis, or the other peculiarities on which the founders of Pteridology once exclusively relied."

The text of the present volumes will, it is hoped, be found little in need of explanation. It may however be briefly mentioned, that besides a full and plain account of the species themselves, an attempt has been made to record, and to give some account of the multitudinous variations which, even in so limited a geographical area as that of Great Britain and Ireland, have been met with by diligent explorers, within a very few years.

Of these varieties or variations, a certain number, of which the characters and synonyms are given immediately following the synonymy of the species itself, are assumed to be of sufficient importance to claim botanical recognition on account of certain marked peculiarities of structure, or the remarkable nature of their monstrous deviations from the normal or type forms.

Besides the above, however, a considerable number of less striking but for the most part permanent and constant variations from the specific types are known to occur. These are included in the complete enumeration of varieties which forms the conclusion of the chapter devoted to the several species; and while it will be apparent, from the subordinate position assigned to them, that no botanical importance is claimed for most of them, yet they are so highly prized by Fern-cultivators, and Fern-admirers generally, as to deserve record for their especial behoof. The rigid scientific botanist or pteridologist may perhaps experience a scientific shudder as he scans the long series of named forms which we have had occasion to record under some of the species; but he must recollect, that if recognised and recorded at all, names are absolutely necessary to prevent general confusion; and recognised they most undoubtedly are by not a few who derive agreeable recreation, either in seeking them amidst enchanting rural scenery, where both mind and body derive benefit from the pursuit, or in tending and preserving them in their ferneries and rock gardens.

It is to be remembered that the variations here spoken of, though sometimes slight, are often in other cases very marked, and are for the most part constant and appreciable. They by no means frequently occur as mere monstrosities, but even when so, they are in general permanent, and renewable from the spores. A considerable number of these varieties have been figured.

We believe, however, that varieties such as these we have recorded, have a botanical significance; that they are, in fact, items in the mass of evidence by which we may arrive at the conclusion that species generally have a wide range of form, even within narrow geographical limits. The variations found among plants, nearly related, though admitted to specific rank, often serve to connect the individuals into a series so extended and withal so complete, that the so-called species themselves seem to lose all definite limit. This surely indicates that these so-called species of plants are but groups of individuals having a certain amount of resemblance, associated by the Naturalist for the mere technical convenience of interchanging information respecting them: just as genera are undoubtedly groups of the so-called species collected together with the same end in view. Nature seems to acknowledge only the individual, while the species (at least such species as are now admitted and are indeed required for the purpose just alluded to,) is an artificial thing of man's contrivance. If indeed, Nature yields *her* vegetable species, as probably she does, for the earth was created to bring forth grass and herb and tree "after his kind," these must be something far more comprehensive than those of the botanist, and perhaps not such as he could employ, at least without subdivision, to carry out his schemes of classification.

Applying, however, the term 'species' to the groups of individuals as usually associated under this designation, whether the association be natural or artificial in character, another consideration arises. If in a small area like that of Great Britain, there occur so many varieties, often marked, and usually constant in character, the variations must become much more numerous and important, if the species is spread, as often happens, over widely separated and extensive portions of the globe. Hence it may be concluded, that very many so-called species, founded on slight differences, and often on the mere fact of geographical separation, are nothing more than local variations. Many apparently genuine species become broken down and amalgamated by the discovery of such connecting varieties, it may be, at their antipodes; and so an extended series of observations may eventually tend to bring the 'species' of the botanist in better agreement with those of Nature's own contrivance. We may at least prove, from the evidence afforded by the variations among the British Ferns, that so-called 'species' do vary very much within a limited area; and may hence draw the inference, sufficiently supported by facts, that they vary much more when a wider range is taken into account; such a conclusion being clearly unfavourable to the multiplication of their numbers.

These considerations do not at all affect the question whether or not the forms, widely different as many of them are, to which we have been referring, are to be recognised and named, as distinct objects. They exist, and they upset all definitions of species when they are ignored; so that in some form or other, whether as 'species' or 'varieties,' they command the attention of even the most rigid and conservative of botanists.

We have only further, in this place, to perform the grateful task of tendering our thanks, generally, to those who have kindly assisted us with materials both for description and illustration. In particular our account of the many variations of some of the species, could not have been so complete without this assistance. We trust that the friendly intercourse now for many years held with the majority of our most enthusiastic Fern-gatherers, may yet long continue.

CHELSEA, *August 31st*, 1859.

CONTENTS.

INTRODUCTION.

THE BRITISH FERNS.

LIST OF PLATES.

INTRODUCTION.

CHAPTER I.

PRELIMINARY REMARKS.

GENERAL FEATURES.—Flowering and Flowerless plants—Distinctions between Ferns and Flowering plants—Distinction between Ferns and other Flowerless plants—Variety of character and aspect—Popular estimation of Ferns—Choice of materials for study.

THE two great classes into which the vegetable kingdom is divided, and to which the terms FLOWERING PLANTS and FLOWERLESS PLANTS are popularly given, are, in the perfect state, readily distinguishable from each other by the peculiar features indicated in the terms thus applied to them. In the one class, flowers followed by seeds, forming the appointed means by which the Almighty fiat, "increase and multiply," is carried out, are developed in due course upon the parent plant. In the other class, there are no parts produced corresponding to flowers; the stem bears leaves only, these leaves having their own varied peculiarities of development, among which, however, is always to be found the property of forming, for the purpose of multiplication, a peculiar form of germ-bud, to which the name of spore is given.

It is to the second of these primary classes, that distinguished by the name of Flowerless Plants, and bearing spores instead of seeds, that the Ferns belong.

Among the secondary groups which, equally with themselves, form part of this great Flowerless class of vegetation, the Ferns are distinguished mainly by the nature and position of the little cases or vesicles, which contain the spores. These vesicular bodies, known as the spore-cases, are placed on the surface or on the margin of the leaves, and consist of small hollow cells without internal divisions or partitions, containing each a great quantity of the spores, which latter are of microscopical proportions, and all possessing the same general structure.

It will thus have become evident that the Ferns are to be known from the great group of Flowering plants, by their not producing either flowers or seeds, properly so called; while they are known from other Flowerless plants by the combination of these characteristics:—(1), their spore-cases are borne on the back or at the edges of their leaves; (2), they are all one-celled; and, (3), the spores they contain are all of one kind.

The Ferns are plants of very varied size and aspect. The majority of them are dwarf terrestrial herbs; many of them are epiphytal; some are minute plants of delicate structure; while others, again, form noble trees with stems of fifty feet or upwards in height, crowned by magnificent plumy heads of fronds. Mr. Ralph states, that in the *Cyathea medullaris*,* one of the tree Ferns of New Zealand, he has counted three dozen of the fronds in full vigour in one of these crowns at one time, the fronds being twelve feet or upwards in length, and requiring considerable effort to raise them when cut off and lying on the ground. Nor is their form or cutting less varied, all gradations of outline being met with from linear to lance-shaped or deltoid, and all intermediate states of division, between the simple or undivided frond, and that in which division and subdivision of the parts is so many times repeated, as to produce a highly composite character.

The elegant characters of outline and subdivision of parts so commonly found in the fronds of Ferns, have led to their being

* Ralph, in *Journal of Proceedings of Linnæan Society*, iii. 166.

associated with all that is graceful and fascinating among vegetable forms, and have gained for them a very high position in popular favour. They are indeed considered the very prototypes of gracefulness. Those who have seen even some of our commonest native species growing in full luxuriance in favoured situations, will have no difficulty in admitting that even they have no mean claim to so exalted a position; and there are numerous exotic species, which no doubt far exceed them in beauty. Gay colours, it is true, are for the most part wanting. In some species, indeed, the surface is frosted with silver or glittering with gold-dust, and the rich warm yellowish and brownish tints of the fructification produce a pleasant contrast to the predominant green, among which, moreover, many varieties of tint are to be found; but on the whole the colouring of Ferns is sober, and their fascinations must be admitted to depend rather on the extreme elegance of their forms, than on their vivid or attractive hues.

The British Ferns are not very numerous in species, but there is a very large number of varieties, and these add greatly to the interest that attaches to the plants. Not only on this account, but by reason of their accessibility, they attract many admirers, and absorb much of the attention of many students, as the interest attaching to the discovery of new forms sufficiently attests. They are, moreover, for the most part very easily cultivated, and beyond all other plants, are perhaps the best adapted to parlour or window culture. To the satisfaction that may be found in the collection and preservation of Ferns in the herbarium, and to the study of them in the dried state, may therefore be added the pleasure derivable from watching their progress in a state of cultivation, and from improving the opportunities thereby afforded of studying their structure, their habits, and their peculiarities, as they become day by day developed in the living state.

Those who desire to form a thorough acquaintance with the

peculiarities of the species of Ferns, should certainly, if possible, adopt the method of studying from the living plant, for it reveals many curious and interesting features which are not to be learned from the most patient and assiduous investigations, carried on by means of such dried portions as are preserved in the herbarium. It is not the less true, however, that all the essential features necessary for the recognition of the species, may be preserved in well-selected herbarium specimens, so that those who have not convenience for cultivating the living plants, may yet store up in their cabinets, for the amusement of their leisure hours, ample materials for acquiring an accurate knowledge of the subject.

When the *Hortus Siccus*, or collection of dried specimens, is depended on for the purpose of study, some judgment should be exercised in the choice of the plants. They should be as nearly perfect as circumstances permit; root, stem, and frond, with their scales or hairs, and their sori should be secured, and none of these should be damaged by careless handling. The fructification, which is most important, should be nearly or quite full-grown, so as to show its mature form and character, but it should be rather under than over ripe, so that, if an indusiate species, the indusium may not have been cast off and lost. In very old fertile fronds the appearance of the fructification often becomes completely changed, and with such materials alone, serious errors of judgment may be committed by those who are not well-grounded in the study.

CHAPTER II.

VEGETATIVE ORGANS OF FERNS.

Acrogenous growth—The ROOTS—Their nature and position—Buttress-like masses of roots in Tree Ferns.—The STEM—Caudex—Rhizome—Their peculiar modes of growth—Accretion—Internal structure—The tissues—Vascular system and venation—The FRONDS or Leaves—Not true leaves—Their diversity—Vernation—Peculiarities of structure—The stipes—The lamina—Different modes of division—Venation—Its peculiarities—Application of terms: costa, veins, venules, veinlets—Arrangement of veins—Venation as a generic feature.

THE Ferns, as already stated, belong to the class of Flowerless plants, and consequently their vegetative organs make up nearly the whole bulk of the plants. They have, in all cases, a very distinct stem which, however, assumes several different characters, as will be hereafter explained; and they are furnished below with roots, or food absorbing organs, and above with leaves of a peculiar character, to which the name of fronds is given.

The group of plants to which the Ferns belong is called Acrogenous; this term being applied to it because the increase in size which takes place as the plants grow, is effected by means of accretion at the apex or growing-point. The several organs of the Ferns are readily distinguishable from each other, by which character they are known from certain other plants of low organisation—the Thallogens, in which the stems and leaves are blended. The vegetative organs of Ferns are the following, namely, the Root, the Stem, and the Frond or Leaf.

The Roots of these plants are always fibrous, and in their younger portions are clothed with fibrils, or fine soft hair-like bodies, which give them a velvety or downy surface. When the stem assumes the form of a rhizome, the roots are produced here and there from its under surface as it creeps along upon or beneath the surface. When, however, the stem is caudiciform, whether erect or decumbent, the roots proceed from among the bases of the old fronds, issuing from the stem on all sides, and sometimes overlying each other in masses. In these cases the fibres or roots are very frequently wiry and rigid in texture, and comparatively coarse. In some of the New Zealand tree Ferns,* *Cyathea medullaris* for example, the fibrous roots which push out from the lower part of the stem closely overlie each other, and form a kind of buttress of variable thickness around its base, extending more or less upwards. This growth sometimes attains a girth of six or seven feet, extending upwards and gradually lessening in thickness to the height of ten feet or more from the ground. In the *Cyathea dealbata*, also, these aërial roots form a wiry fibrous mass exterior to the true stem, gradually enlarging so as to become about a foot and a-half in diameter towards the base. The tendency of the rootlets is usually downwards, but sometimes, as at the upper part of the crown of *Osmunda regalis*, they may be observed shooting out in all directions, some of them being directed upwards. The root-fibres often form entangled masses of considerable size. The fibres are sometimes simple, oftener branched, and generally, as before remarked, velvety or downy on the younger portions. These organs, of course, act as absorbents to supply the plants with nutriment, which nutriment is obtained from the soil in the case of the terrestrial species, and more or less from the atmosphere in those which are of epiphytal habit.

The Stem of the Fern, often erroneously in popular language called the root, assumes two distinct forms, the peculiarities of which have been employed by Mr. J. Smith as primary characters in a new mode of classifying Ferns, which he has proposed.† These

* Ralph, in *Journal of Proceedings of the Linnæan Society*, iii. 165.

† J. Smith, *Botany of the Voyage of H.M.S. Herald*, 226.

two forms are those of a *caudex* and a *rhizome*. In the caudex or caudiciform stem, the fronds rise from the termination of the axis of growth, either in a single series, or in a kind of crowded whorl, so as to form a terminal crown. The young fronds in all cases spring from the inner side of those previously formed, their bases becoming united and adherent with the axis, so that the older part of the stem consists of a combination of the axis with the bases of the fronds developed from it. This manner of growth is exemplified in the common Male Fern, and in those species which form the tree Ferns of the tropical forests. In the rhizome or rhizomiform stem, the fronds, which are more or less scattered, are developed from the sides of the axis of growth, the point of which is *apparently* in advance of the last formed fronds; they are, moreover, developed from nodes formed on the rhizome, each node producing a single frond, which has an articulation or joining at or near to the base of its stipes, at which point it eventually separates spontaneously. The common Polypody affords a good illustration of this mode of development.

The caudiciform stem or caudex, is in numerous instances among the dwarf herbaceous species scarcely at all lengthened, but erect and tufted, forming a cæspitose series of crowns, whence the fronds issue. Even among the species of herbaceous habit, however, the caudiciform stem occasionally in age becomes considerably elongated, and such instances, which may be found among our native species, afford an imperfect and pigmy illustration of the manner in which the trunks of the tropical tree Ferns are formed. This elongation of the older caudices may be observed frequently in the *Osmunda regalis*, which among our native Ferns is that which most frequently assumes this habit, being sometimes found with bare stems, a foot or more in height. The same tendency, though in a less degree, may be observed in very old plants of the *Lastrea Filix-mas*, *Lastrea montana*, and *Lastrea dilatata*, and sometimes also in *Athyrium Filix-fœmina*, and *Polystichum angulare*.

The erect stems of Ferns vary much in bulk, and in some species branch freely, producing many tufted crowns, but in other species they are very rarely at all divided. It is no doubt partly to this tendency to development only from the terminal bud, that the elevation acquired by the stems of tree Ferns is due.

Various terms are employed to express the peculiarities which occur in this caudiciform stem. Thus when it is simple and erect, and acquires an elevation of about two or three feet or upwards, it is said to be *arborescent.* The dwarfer forms of upright-growing simple stems are generally described as *erect,* the fact of their being dwarf and not arborescent, being understood. The dwarf caudiciform stems are not, however, always erect; they sometimes grow horizontally along the surface of the ground, and are then *decumbent,* or if much elongated, with the fronds distantly placed, *creeping* or *scandent.* The decumbent form of caudex would be well represented by a stem of *Lastrea Filix-mas* which had fallen aside to a horizontal position; while the creeping caudex is illustrated by that of *Lastrea Thelypteris,* or *Pteris aquilina,* the latter not only extending to a considerable distance, but penetrating to a considerable depth. The wiry stems of *Trichomanes,* and the fine thread-like stems of *Hymenophyllum,* seem to be rather modifications of the creeping caudex, than rhizomiform stems as they are often considered. When the stem is very short and branched so that several crowns are formed in a cluster, it is said to be *tufted;* this is illustrated in the *Cystopteris fragilis.*

The rhizome is, in almost all cases, of creeping or scandent habit, and is very frequently epiphytal. In some exotic genera, as for example in *Oleandra,* it is, however, erect, and produces roots from its lower parts, on all sides indifferently. In this form of stem, the bases of the fronds are not continuous with the axis, as in the case of the caudex, but they have a natural joining or articulation at which they separate spontaneously when their functions are completed, leaving a concave cicatrix or scar. When the stems assume this creeping habit, they usually if terrestrial extend either on the surface, or just below the surface of the earth, often becoming branched, the fronds springing up individually and distinct, and more or less widely separated. If epiphytal they creep along the surface of the bark of the trees that support them, or hang dependent from their branches; they are sometimes as thick as one's finger, as in the *Polypodium vulgare*, but are often much smaller than this. The true rhizomes are generally thickly covered with scales, which are variable in size and form, and are sometimes so large and numerous as to render the surface quite shaggy. Unless scales are common to the whole frond, they seldom extend in Ferns of this habit beyond the point at which the fronds are articulated.

The rhizome or creeping stem, where it exists, affords great facilities for propagation; for if a portion of moderate length, furnished with fronds and roots, is separated from the rest, and placed under proper conditions, it will readily form an independent plant.

Whether the stem is caudiciform or rhizomiform, erect or creeping, accretion takes place, according to Hofmeister, by means of a continued multiplication of one apical cell.

As regards their internal structure, the stems of Ferns are more highly developed than those of other acrogenous plants. The lower groups of these Acrogens consist merely of cellular tissue, but in the more highly organised forms, among which the Ferns take precedence, both woody and vascular tissues are found. In some Ferns, *Marattia* for instance, the vascular bundles are regularly distributed throughout the whole mass forming the stem. In others, such as the *Lastrea Filix-mas*, they are disposed in a single circle, or, as in *Pleopeltis leiorhiza*, with neighbouring smaller bundles. In others, again, as in *Trichomanes reniforme*, they are reduced to one central bundle; while instances occur in which they are disposed on either side of hard plates, as in *Pteris aquilina*. In many of the tree Ferns, a few bundles larger than the rest, are closely surrounded with dense tissue, and disposed symmetrically around the axis, sometimes forming a close cylinder, as in *Dicksonia antarctica.* Mr. Berkeley mentions* a tree Fern from Sylhet, in which this cylinder was found to be nearly perfect. In some cases, Mr. Berkeley continues, they seem to be quite isolated, giving off no bundles to the fronds, this office being performed by smaller fascicles, as in the tree Ferns figured by Mohl; † while in others, they as evidently supply the stipites.

The general disposition of the tissues in the more highly organised forms of Ferns, has been described as follows: ‡—Round the scars of the stipites there are cavities filled with stellate brown tissue. The cortical stratum consists first of cuticle, then of parenchym, and then a harder layer of brown sometimes parenchymatous sometimes prosenchymatous tissue with thick punctated walls. The enclosed cylinder is filled with softer cellular tissue, containing many cysts gorged with resinous matter, and various bundles of vascular tissue, attended by pale pleurenchym. The larger bundles which are flattened and variously curved, are surrounded by dense tissue like the inner layer of the bark, and arranged in a circle symmetrically round the axis, with short interspaces, through which the other smaller bundles dispersed in the central mass give off branches into the stipites, while others exist in the space between the bark and larger masses. These latter form a cylinder, more or less perfect, round the axis, and are altogether distinct from anything in endogenous stems, besides which there is not that crossing of the fascicles characteristic of Endogens. The vessels Mr. Berkeley states are always scalariform; but Dr. Lawson informs us, that he has frequently found unrolled spiral vessels.

* Berkeley, *Introduction to Cryptogamic Botany*, 514 (note).

† Martius, *Icones Selectæ Plantarum Cryptogamicarum Brasiliensium*, tt. 29-36.

‡ Berkeley, *Introduction to Cryptogamic Botany*, 514.

The vessels, Mr. Berkeley continues, though varying greatly in size, are accompanied by cellular tissue, and surrounded by elongated pale wood cells (pleurenchym), beyond which is parenchym, mixed with resinous cysts. The hard coat which encloses the vessels with the pleurenchym and parenchym, belongs to the general mass of cellular tissue, and not to the wood, which is represented by the minute quantity of pleurenchym.

The structure of other Ferns corresponds more or less closely with this, the chief difference being in the disposition of the bundles and of the hard attendant tissue, which is infinitely varied. Thus in *Pteris aquilina*, the hard tissue which in the tree Ferns encloses the principal vascular bundles, is disposed in two curves on either side of which the vascular bundles are arranged, each enclosed by a thin coating of tissue denser than the general mass. Sometimes these two arcs meet at one extremity, sometimes at both, especially when a stipes is given off. Besides these two main masses of dense tissue, there are many scattered fibres. In other cases, this peculiar dense tissue is altogether scattered about in little fibres, like those of *Pteris aquilina* just mentioned; in some, as in *Oleandra hirtella*, it is either converted into or replaced by curious cysts which display a fibrous structure. In many cases, as in *Pleopeltis leiorhiza*, where the bundles are arranged in one principal circle with a few outliers, it appears to be entirely wanting. In this, and many other Ferns there appears to be scarcely any distinct cortical layer, the transition to the general mass of cellular tissue being almost imperceptible.*

The same kinds of tissues are found in the stipites and rachides, where there occur detached rounded bundles, or flattened, often curved, plates of the harder tissues, in the midst of the cellular mass. The arrangement and number of these bundles and plates

* Berkeley, *Introduction to Cryptogamic Botany*, 515.

have been resorted to as affording generic or specific distinctions; but for this purpose they have little value, for they are variable both in number and arrangement in plants of the same species of different degrees of vigour, as well as at different elevations of the same stipes. The system of vessels is further carried through the frond, in the ribs and veins which traverse the various leafy parts into which it is divided. It is this connection of the veins of the fronds with the entire vascular system, and the fact that it is on some part of these veins that the spore cases are borne, which gives to venation its value in the systematical arrangement of these plants.

The Fronds of Ferns are the leaf-like organs which are borne on the proper stem. The leaf-like character they bear, has led some botanists to consider them as true leaves, and to reject the term frond altogether. This term—frond, has however long been, and still is commonly used for that part in the Fern which occupies the place of the leaf in flowering plants, and it is conveniently retained to express the fact that these two organs are not quite identical in character. The frond of the Fern differs from the leaf of the flowering plant in this, that it actually bears on its surface the parts known as the fructification, which the true leaf does not. This is the popular view. In an exact physiological sense, perhaps, this difference does not exist, inasmuch as the so-called fructification of Ferns is not strictly analogous to the fructification of flowering plants, but is rather to be considered as analogous to the little buds or bulbils which some plants throw out from their surface. The peculiar manner in which the reproductive parts are borne on this portion of the plant may, however, be taken to indicate an essential difference, or at least to show that the frond is something more than the leaf. An analogy has been traced between the fronds of Ferns and the deciduous branches of other plants; but neither are they, properly, branches, and the analogy does not hold good, because, though fronds are sometimes articulated with the stem, especially in the case of those of creeping habit, yet they are not so always.

It is in the fronds that we must seek for that ornamental character which renders the Ferns so popular. The fronds alone, however, afford an almost endless variety. Some of them are very large, others are very small; some are quite simple and not at all divided, while others are divided beyond computation into little portions or segments. It is these much-divided fronds which are generally regarded as the most elegant. Even in the few species which are natives of Britain, this variety of size and form is very obvious. There are some among our native species which are not more than two or three inches high, while others are five to six feet or more in height; and they vary from less than an inch to two feet or more in width, presenting great differences of outline, of which, however, the most common are: the lance-shaped, the ovate, the triangular or deltoid, and the pentangular. Some fronds are quite simple, and others are cut into innumerable small segments. There is variety of texture, too, among the Ferns; some being so thin and delicate as to be almost transparent, while others are thick and leathery, and some perfectly rigid. And in respect to colour, while green predominates, some are pale green, some deep green, some have a peculiar glaucous or sea-green tint, and some are dark brownish or olive green. The surface is variously smooth and shining, or opaque, or less frequently, covered with hair-like scales.

The fronds of almost all Ferns are, in their incipient condition, coiled up inwards towards the axis of development, forming a series of convoluted curves. This folding up of the fronds of Ferns, as of the leaves of other plants, is termed their *vernation;* and this peculiar form of vernation is that called circinate, or *gyrate.* The only British Ferns which have not circinate vernation, are the *Botrychium* and *Ophioglossum,* in which the parts, instead of being rolled up while undeveloped, are simply folded together, the vernation being *plicate*. The more compound of the circinate species have not only the frond as a whole, but its divisions also, rolled up in a similar manner; in such cases, the larger divisions first open, and the rest follow in succession. In many species, the fronds, when partially developed, have a very peculiar and graceful appearance.

When the fronds become fully developed, two parts, the *Stipes* and the *Lamina*, are distinguishable. The stipes is the stalk, the lamina is the dilated leafy expansion.

The stipes or stalk, sometimes by error called the stem, which latter term properly belongs to the caudex, is formed of a hard, external layer, covered by a cuticle, and enclosing a mass of cellular tissue, traversed by plates or bundles of vascular tissue, disposed in some regular order. The number and position of these vascular bundles have been considered sufficiently constant and important to afford characters for discriminating genera and species, but for this purpose they are valueless, as they differ in the same plant at different elevations of the stipes, as well as in stipites of different degrees of vigour at points of equal elevation. The lower part of the stipes, generally, and sometimes even the entire length of the *Rachis* —which is the continuation of the stipes through the leafy portion of the frond, is more or less furnished with paleaceous or membranous scales. These scales, which are generally brownish, are in some cases confined to a few small bodies scattered sparingly near the base of the stipes, but in other instances are so large and numerous as to produce a shaggy surface. They are, no doubt, appendages of the same nature as the hairs and scales found on the surface of other plants. Their form, as well as number and position, and even colour, are found to be tolerably constant in the different species or varieties, and hence they sometimes afford marks of recognition. Whenever they are produced along the rachis, as well as on the stipes, they are invariably largest at the base, and become gradually smaller upwards. In most of the rhizomatous Ferns, the base of the stipes is articulated with the stem, that is to say, it is furnished with a natural joint or interruption of the woody fibres, so that in age it separates spontaneously. This is not frequently the case with the caudiciform kinds. When the firmer tissues of the stipes are continuous at the base with those of the caudex, so that the fronds do not separate spontaneously, the fronds, or stipites, are said to be *adherent to the stem;* but when there is a natural articulation or joining, so that, when its functions cease, the frond separates and falls away, leaving a clean scar on the rhizome or caudex, the fronds or stipites are said to be *articulated with the stem.* The continuation of the stipes upwards constitutes the costa or rachis, to which further allusion will be made. The stipes and rachis may be either green, or of some distinct colour; if the latter, it is usually a dark purplish brown, or blackish purple.

The upper leafy portion, or lamina of the frond, is that to which the name of frond is most frequently and especially given, irrespective of the stipes, which really forms part of it. This part affords great variety in the mode in which it is divided. A certain and by no means inconsiderable number of Ferns, have undivided fronds; these are called *simple.* When they are partially once divided they are *pinnatifid,* and when quite once divided, that is divided down to the rachis, they are *pinnate.* When partially twice divided they are *bipinnatifid,* and when thoroughly twice divided, *bipinnate.* So, when partially or thoroughly thrice divided, they are *tripinnatifid* or *tripinnate;* or, if four times divided, *quadripinnatifid* and *quadripinnate.* Many Ferns are once pinnate and then pinnatifid; such a mode of division is called *pinnato-pinnatifid.* The divisions of the pinnatifid frond are called *segments;* those of the pinnate frond, *pinnæ.* In the bipinnate frond, the first series of divisions are pinnæ, the second *pinnules,* or little pinnæ; in the tripinnate frond, the first series are *pinnæ,* the second series *primary pinnules,* the third series *secondary pinnules,* and so on with the rest. The correct appreciation of these terms is necessary to the comprehension of descriptions of the plants. The peculiarities of the division of the frond indicated by those terms, are much employed in distinguishing the species of Ferns, and are moderately constant and reliable.

The parts of pinnated or bipinnated fronds are, in some species, joined to their respective rachides by a natural joint, as the stipes is, in some species to the stem, the firmer tissues not being continuous at that point, so that the parts, whether pinnæ or pinnules, readily detach themselves after maturity is past. When so jointed the parts are said to be *articulated with the rachis.*

In the majority of Ferns, the mature fronds are alike fertile, and are similar in their appearance; but in certain species, the habit is to produce some of the fronds wholly barren, and others wholly fertile; in these instances the fertile fronds are more or less contracted. This habit is exemplified, among our native species, in the *Allosorus crispus*, and *Blechnum Spicant.*

The fronds of Ferns are variable in their duration like the leaves of other plants. In some species they are persistent, so that either absolutely, or with very slight shelter, the plants become evergreen. The species possessing this habit are the most valuable for the cultivator. In other species the fronds are fragile, and of short duration, produced only during the warmer portion of the year, and shrinking before the first breath of winter. Among these latter, however, are comprised some of the most delicately beautiful species.

Fern fronds are traversed by ribs or fibres, which serve to give them their elasticity. These ribs are what are called the *Veins.* The venation thus forms a framework, on which the herbaceous portions are, as it were, spread out to the influences of light and air; and, consisting of bundles of woody fibre traversed by the nourishing vessels, it constitutes the vascular system of the plant. The stipes mainly consists of this vascular tissue, definitely arranged, and this is continued onwards throughout the leafy part of the frond, becoming branched more or less according as the frond itself is more or less divided, the branches, according to their position, forming the rachis, costa, veins, venules, or veinlets.

When the frond is simple, that is undivided, as in the *Scolopendrium vulgare*, the rib or continuation of vascular tissue from the stipes through the frond, forms the *Costa* or midrib. From the costa are given off, laterally, branches more slender than itself, of which the first series are called *Veins*; if these veins are branched, the branches form the *Venules*, and if the venules are again branched, their branches are *Veinlets*—the last series, whenever the venation is very compound, being distinguished as the *ultimate veinlets*. If the frond is pinnate, that is, divided into separate leaflets, or *Pinnæ*, as in *Asplenium marinum*, the analogue of the rib, which in the simple frond formed a costa, then becomes the *Rachis*, and each separate pinna has its own costa and series of veins as the undivided frond had. If the frond is bipinnate, that is, having the pinnæ again divided into separate leaflets called *Pinnules*, the part corresponding with the costa of the undivided frond becomes the primary rachis, or rachis of the frond; and that which corresponds with the costa of the pinna in the once-divided frond, becomes the secondary rachis, or rachis of the pinna; the pinnules each acquiring their respective costa and veins. If the frond is tripinnate, that is, having the pinnules distinctly divided, it then acquires a tertiary rachis and secondary pinnules, or *Pinnulets*, and these latter are furnished as before with a costa, and the series of veins proceeding from it. Once more divided, the fronds become *quadripinnate*, and these and such as are still more divided are called *decompound*. The several series of pinnæ and pinnules in compound fronds may in any case be distinguished as primary or secondary, &c.; and the last series of divisions in a very much divided frond, are most readily and distinctly referred to as the ultimate pinnules or segments, according as they may be either entirely or only partially separated.

It will now be seen, and it is a point important to observe, that the costa and rachis of a simple frond, are not the same parts as those of a pinnate, or of a bipinnate frond, but that the application of the terms changes according to the degree in which the frond is divided. The costa is the principal rib of the simple frond, or of the last series of distinctly formed leaflets in divided fronds; while the rachis in a divided frond is the part answering to the costa in the simple frond, that is to say, the direct continuation of the stipes through the lamina. The position of the costa once settled, the first branches from it are always the *veins*, the branches from these are the *venules*, and the branches from the latter, the *veinlets*. The foregoing statements respecting the application of the term costa, have been made to apply to those fronds which are either simple or distinctly divided into separate leaflets. It has been already mentioned, however, that there is another mode of division intermediate in character between these, called pinnatifid, bipinnatifid, tripinnatifid, &c., the divisions in these cases not being separate leaflets, but lobes, more or less deeply separated, of the scolloped margin. This does not affect the position of the costa, unless the lobes are very shallow indeed, so that in general the midrib of the lobe would also be called the costa.

The venation of the frond, which has now been shown to mean the costa and its ramifications, presents great variety in different species, but it is, in a general way, quite constant in the same species. There are a few instances known, in which, for example, Ferns which usually have the veins free, have them here and there running together in particular fronds or parts of the fronds, but these are to be regarded as mere exceptions which prove the general rule, and are not often liable to be misunderstood. Indeed this inconstancy most generally happens in cases where some disturbance of the ordinary development has occurred. Thus in the *Scolopendrium vulgare*, which normally has entire fronds and free veins, some of the monstrous varieties, in which the frond is narrowed and the margin broken up, have the veins so entirely disarranged, that they unite more or less frequently. We believe that all the instances which have been observed of inconstancy in the arrangement of the veins, may be referred either to an accidental confluence, or to some obvious disturbance of the normal development.

The arrangement of the veins in Ferns may be brought under four heads, as follows:—(1), *free*, i. e., where the veins are continued without coming in contact with each other; (2), *connivent*, i. e., where all having an excurrent or outward tendency, they coalesce anglewise; (3), *combined*, i. e., where a longitudinal vein unites continuously and transversely with other veins; and (4), *reticulated*, i. e., where they form a complete network, being united in every direction.

With the exception of *Ceterach officinarum* and *Ophioglossum vulgatum*, which have their veins reticulated, all British Ferns belong to the division with the veins free. There are one or two peculiarities still to mention. In some Ferns the venation is not developed beyond the first series from the costa—the veins, these remaining unbranched. In others, the costa itself is not present, but the whole venation consists of forked veins, venules and veinlets, branching in a flabellate or fan-shaped manner from the base of the pinna or pinnule; this occurs in *Adiantum Capillus-veneris*, and in *Botrychium Lunaria*. There is, moreover, to be observed a difference in the mode in which the branching of the free veins is effected; some being pinnate or feather-branched, as may be well seen in the larger varieties of *Lastrea Filix-mas*, while others are forked or dichotomously-branched, as in *Scolopendrium vulgare*.

It is, as we have elsewhere observed, the condition of the veins and the mode of their arrangement which has of late years been employed as an auxiliary character in distinguishing the genera of Ferns. This use of the venation has been objected to by some botanists of high authority, who think that the fructification alone should furnish the generic character. Such a rule may be quite proper as regards flowering plants, where the organs forming the parts of fructification are numerous and varied, but it does not appear to be equally so in the case of Ferns, where the so-called organs of fructification present few available differences, and where the species are nevertheless so numerous, that further distinguishing characters are desirable in order to break up into groups of moderate size the unwieldy genera of olden times. No auxiliary character that has yet been suggested, has proved so useful, nor so constant, nor as we believe, so important, as the venation. That such is the case, is indeed practically admitted, even by those who object to its use in distinguishing genera, for they willingly employ it as a characteristic of sub-genera. Now the difference between a genus and a sub-genus is so very trifling, the limitation of genera being a mere matter of fancy or convenience, and therefore varying in the estimation of different persons, that the admission of venation as a characteristic of the lesser conventional group—the sub-genus, may be taken as admitting it in the case of the larger but equally conventional group—the genus. Much of the importance we are inclined to claim for venation as a feature of generic value, rests upon the fact that IT gives rise to the fructification. The spore-cases which form this fructification spring out of it; the receptacle to which they are attached is part of it; and this intimate connection with the fructification must give the venation in the Fern a higher importance than could be properly attached to it in the case of flowering plants, though even among the latter, the important differences of free and reticulated venation run so exactly parallel with other features which mark the great primary groups of Exogens and Endogens, that they are the *primâ facie* characters universally employed in distinguishing them.

CHAPTER III.

REPRODUCTIVE ORGANS OF FERNS.

The SORI—Their form and position—Dorsiferous and Marginal-fruited Ferns.—The RECEPTACLE—its systematic value—its position.—The SPORE-CASES—Annulate and Exannulate, or Ringed and Ringless Ferns—Vertical and Oblique rings—Rudimentary ring—Indusium—Involucre—Indusiate and Non-indusiate Ferns.—The SPORES—Their germination—Prothallus—Antheridia—Pistillidia or Archegonia.

THE Reproductive Organs of Ferns consist of *spores,* enclosed in *spore-cases,* sometimes called *sporangia, thecæ,* or *capsules,* these spore-cases being collected into groups of varied form, called *sori* (sing. *sorus*).

The Sori consist of the spore-cases, collected together into groups of various outline, usually forming distinct spots, round or oblong, or lines more or less extended, and occupying different positions with respect to the venation. They are generally situated on the plane of the frond, or slightly elevated, and are then called *superficial;* but sometimes they are seated in a groove or channel, and they are then said to be *immersed.* The principal forms assumed by the sorus are:—the *punctiform,* i. e., forming round dot-like clusters; the *oblong* or *linear,* i. e., forming lines, short or long, which lines may be either marginal, costal, or oblique, and either simple or united like net-work, following the veins; the *amorphous,* i. e., forming large shapeless patches; and the *universal,* i. e., covering the whole surface.

In the annulate Ferns, which comprise all the *Polypodiaceæ,* the sori are usually borne on the under surface, or, what is usually called the back of the frond. The sori are then said to be *dorsal,* and Ferns which have their fructification in this position are called *Dorsiferous.* Most of the British Ferns are dorsiferous. When the fructification is protruded from the edge of the frond, it is said to be *extra-marginal.* In the extra-marginal fruited group the spore-cases are often collected around the free extremities of the veins, which are surrounded by thin urn-like expansions of the cellular tissue, as in *Trichomanes* and *Hymenophyllum.* In *Osmunda* the marginal spore-cases are quite exposed.

The normal development of the dorsiferous Ferns is, of course, to bear their fructification on the under surface or back of their fronds, but in this dorsiferous class there are some curious deviations from the rule, the sori growing either above instead of beneath, or above as well as beneath; while in some instances those having their fructification normally extra-marginal, have also produced sori on the plane surface of the frond, both above and beneath. The first noted instance of this deviation from the normal state, of which we have any knowledge, was recorded by us* in the early part of 1856, in the folio edition of this work. We then had occasion to mention that certain varieties of *Scolopendrium vulgare* habitually produce sori on the upper as well as on the under surface of their fronds. This occurs, for the most part, in those varieties, several in number, in which the margin is deeply lobed. In these cases it often appears as if the normally-placed sori, opposite the acute sinuses of the lobes, had been continued to the margin, and then returned on the upper surface; though in many instances the abnormally-placed sori are distinctly within the margin, and borne in positions where there are no corresponding sori beneath. Another instance of this abnormal position of the sori, occurs in the *Polypodium anomalum,* subsequently recorded and figured by Sir W. J. Hooker.† In this, which is a plant of polystichoid aspect, the sori are mostly

* Moore, *Ferns of Great Britain and Ireland, Nature-Printed,* t. 42.

† Hooker, *Kew Journal of Botany,* viii. 360, t. 11.

on the upper surface, very rarely indeed on the under. Sir W. J. Hooker also mentions* an Italian specimen of *Asplenium Trichomanes*, which had produced a solitary perfect sorus on the upper surface of one of its pinnæ; and the same example is alluded to by Mr. Berkeley.†

The above are instances of strictly dorsiferous Ferns becoming more or less supra-soriferous; but we have since had occasion to record ‡ a different kind of aberration, that of an extra-marginal-fruited Fern becoming at the same time both dorsiferous and supra-soriferous. Specimens of *Cionidium Moorii*, grown at Sydney, exhibit this peculiarity. In *Cionidium*, as in *Deparia*, from which latter *Cionidium* is an offshoot with reticulated venation, the sori are normally placed in little cup-shaped involucres set along the extreme margin of the frond, and attached by little footstalks. In the abnormal specimens referred to, such marginal exserted sori were abundant, but besides them, other sori were scattered here and there, both on the upper and under plane surface, entirely removed from the margin, and in several instances placed near the midrib. These aberrant sori were considerably more numerous above than beneath, and were quite perfect, except that they had no appreciable stalk.

The point at which the sorus is fixed to the frond is called the *Receptacle*. The receptacle is formed by an expansion of the tissue at some particular and determinate point of the venation, sufficiently constant to acquire systematic importance from this very fact. In one group only does the fructifying power seem to be more diffused, and here the whole surface is converted into a receptacle; but in the vast majority of cases the sori are placed on some definite point of the venation. Thus a very close connection exists between the venation and the fructification; and in consequence of this, the venation has, we think, been very properly used freely for the

* Hooker, *Kew Journal of Botany*, viii. 361.

† Berkeley, *Introduction to Cryptogamic Botany*, 509.

‡ Moore, in *Journal of Proceedings of the Linnæan Society*, ii. 129.

purpose of affording discriminating generic characters in some modern systems of classification.

The nature and position of the receptacle appears to us, indeed, to be, theoretically, of considerable importance in a systematic point of view, as being the point of contact between the vascular and reproductive systems of the plants. Practically, it does prove of some importance in cases where the fructification is otherwise not easily made out. Thus, for example, in the genus *Cheilanthes* the indusium is often linear and continuous, quite like that of *Pteris*, the spore-cases by confluence often appearing to be continuous also; consequently, unless the actual attachment of the spore-cases is taken into account, there is no determinate limit between these genera. But in *Pteris* proper there is a linear continuous receptacle, transverse to the general course of the veins, while in *Cheilanthes* the receptacles are punctiform at the tips of the veins. The sori are consequently not really, but only apparently linear in the genuine species of *Chielanthes*, and they are thus widely separated from *Pteris*, which otherwise they seem to resemble. This being so, the interpolated genus *Pellæa* is unnecessary. So in *Platyloma*, the sorus is a linear mass of spore-cases, broad indeed, but apparently at the margin, and with what appears to be a narrow marginal indusium. The species are hence sometimes mistakenly referred to *Pteris*, and they are, indeed, generally classed in the group of *Pterideæ*. But instead of having the linear transverse marginal receptacles of *Pteris*, the receptacles of *Platyloma* are really set in lines the other way, that is, extending from the point of each vein more or less downwards. It is only in consequence of their being near together, so that the lines of spore-cases become confluent laterally, that they look at all like *Pteris*, with which they really have no structural affinity. They are, in fact, much nearer the *Adianteæ*, only there the receptacles are resupinate, that is, on the reflexed indusiate portion, and therefore in reality turned upside down.

The position of the receptacle on the vein is variable, although generally uniform in the same species. Sometimes it occurs at the apex of the vein, and it is then said to be *terminal.* Less frequently it is at the point where the vein leaves the costa, and it is then said to be *basal.* Sometimes it is placed at a point between the apex and the base, when it is *medial.* It is sometimes transversely continuous, combining, as it were, the adjacent veins crosswise. Sometimes it occurs at a point where two or more veins unite, when it is called *compital,* but there are no examples of this structure among British Ferns.

The spore-cases of Polypodiaceous Ferns, which group includes the greater number of known species, are small roundish or obovate, hollow laterally compressed one-celled bodies, nearly surrounded by an elastic belt or *Ring,* which ring is sometimes called the *Annulus,* and hence the Ferns of this group are sometimes named the *Annulate Ferns.* In the Marattiaceous and Ophioglossaceous Ferns, the spore-cases are mere one-celled cavities without any trace of ring or annulus, and these plants are consequently called *Exannulate Ferns.*

The exannulate Ferns, which comprise only the *Marattiaceæ* and the *Ophioglossaceæ,* have their spore-cases dorsal on normal fronds in the former group, but collected upon the sides or surface of contracted fronds in the latter. The only exannulate British Ferns are the *Botrychium Lunaria* and the *Ophioglossums.*

The ring of the spore-case takes a different direction in different groups of Ferns, but always forms a single line, partly or entirely encircling the spore-case, and always occupying a similar position in Ferns of the same group. The position and nature of the ring have, in consequence been taken as affording the best technical characters for the purpose of arranging Ferns into primary groups. Some of the annulate Ferns have the ring in a rudimentary condition, but in the majority of British species belonging to the *Polypodineæ,* a division of the *Polypodiaceæ,* it is nearly complete, and occupies a vertical position, extending from the hinder base of the spore-case up the back and over to the front side, extending there some distance downwards. The basal part of the ring is also often lengthened out into the form of a pedicel or stalk, by which the case is attached to the receptacle, though sometimes hardly any elongation takes place, beyond the base of the cases, which are then sessile on the receptacle. The spore-case itself consists of a thin cellular shell, without internal divisions, traversed externally by a single line of short transverse parallel thickened cells, which form the belt or ring. It seems to be the elasticity of this ring which causes the rupture of the enclosing case, and the liberation of the spores, for if examined at the mature stage, especially when exposed to a drying atmosphere, they may be seen to burst open with a sudden jerk, which scatters the spores in all directions. After this rupture has taken place, the spore-cases may be seen to consist of two hollow helmet-like portions, held together by the recurved ring, which looks like a jointed strap. This rupture of the spore-cases in the case of the vertical-ringed Ferns, takes place by an irregular transverse fissure at a point on the anterior side, which M. Fée calls the *stoma* or mouth, where the striæ or joints of the ring become dilated into elongate parallel cells, the weakening incidental to which elongation no doubt facilitates disruption. The bursting of the spore-cases takes place with considerable force, but the two parts into which they become split asunder, are held together by the ring itself, which does not separate at that part which was opposite the stoma in the perfect spore-case.

In the *Trichomanineæ*, which includes the British genera *Trichomanes* and *Hymenophyllum*, the ring of the spore-case is horizontal or oblique instead of vertical, and the spore-cases are vertically compressed. In cases where this structure occurs, the rupture or fissure of the ring is vertical.

In the *Osmundineæ* the ring is reduced to a rudimentary condition, being represented by a few parallel cells obliquely placed near the apex of the spore-case, which bursts vertically across the apex into two nearly equal and regular parts, or valves, the rupture, though extending much further than the parallel cells just mentioned, being no doubt facilitated by their presence.

The rest of the British Ferns belong to the *Ophioglossaceæ*, an exannulate group. Here the spore-cases are quite sessile, and open by regular valves. In *Ophioglossum* there is no spore-case beyond the hollowed substance of the contracted spore-bearing leaf.

In certain of the annulate groups of Ferns, represented by *Polypodium* itself, the spore-cases spring from the surface of the fronds without any perceptible covering; and in these cases the sori are said to be *naked*. In some other groups they are covered while in a young state by a membrane of the same form as their own, which membrane at length bursts according to its natural habit, and is either cast off or pushed aside, as the spore-cases increase in size; this membrane, called the *Indusium*, is very well represented in the young fructification of *Lastrea* and of *Asplenium*. The membranous cover is usually placed above the spore-cases, which is the true position of an indusium; but in one or two comparatively limited groups of genera, a similar kind of membrane is placed *beneath* the sorus, and continued in an incurved manner, so as to invest it; this particular form, which has been called an *Involucre*, is only found among the British genera, in *Woodsia*. From these peculiarities of the cover or indusium the annulate Ferns have been subdivided into two lesser groups, called the *Indusiate* or covered, of which the *Involucrate* is a modification, and the *Non-indusiate* or naked Ferns.

The spore-cases of Ferns seem to have been long considered as special organs, not having any very clear analogy with any part of the structure of Flowering plants. Dr. Lindley, however,* has

* Lindley, *Vegetable Kingdom*, 75.

suggested that they may be considered as minute leaves, having the same gyrate mode of development as the ordinary leaves or fronds; their stalk being the petiole, the annulus the midrib, and the case itself the lamina with the edges united. This view appears to have originated in a persuasion that there was no special organ in Ferns to perform a function which in flowering plants is executed by modifications of the leaves, and also in observations made on viviparous species. It has often been shown that the leaves of flowering plants have the power of producing leaf-buds from their margin, or from any point of their surface. In Ferns, which are exceedingly subject to become viviparous, the young plants often grow from the same places as the spore-cases or from the margin, and they have even been observed to form little clusters of leaves in the place of sori. In these young plants the more perfect though minute leaves are preceded by still more minute primordial leaves or scales, the cellular tissue of which has nearly the same arrangement as the cellules of the spore-cases, and the midrib of which has a striking resemblance to the ring. The above theory, however, is applied only to the gyrate vertical-ringed Ferns. In those which are furnished with a transverse ring, it is suggested that either the midrib of the young scale, out of which the case is formed, is not so much developed; or the case is a nucleus of cellular tissue, separating both from that which surrounds it, and from its internal substance, which latter assumes the form of spores, in the same way as the internal tissue of an anther separates from the valves under the form of pollen.

The spores are minute, roundish, angular, or oblong vesicles, consisting of two outer layers, or coatings, enclosing a thickish granular fluid, and they are very numerous and arranged without order within the spore-cases. They are so small and dust-like, that when thinly scattered over a sheet of paper they are scarcely visible to the naked eye, though lying by thousands amongst the also minute emptied spore-cases. The colour, no less than the form of these spores, is variable; they are usually pale brownish or yellowish, but they are sometimes green, and the tints of brown and yellow are much varied. These organs differ obviously from seeds, in that they consist merely of a homogeneous cellular mass. In true seeds the radicle or young root, and the plumule or young shoot, are present in the embryo, and are developed from determinate points; but Fern spores, consisting merely of a small vesicle of cellular tissue—a vegetable cell, grow indifferently from any part of their surface, the parent cell becoming divided into others, which are again multiplied and enlarged, until a small green germinal scale, or primordial frond, is formed, and from this, in due time, the proper fronds are produced. The surface of the spores is sometimes smooth, sometimes tuberculated, or even echinate.

The germination of the spores of Ferns has lately excited much inquiry, the result of which leads to the inference that something like sexuality exists among all the higher groups of the Cryptogamous plants, a kind of fertilisation taking place on the *prothallus* or germ-frond, which in the Ferns, as already mentioned, takes the form of a small leafy scale. The balance of evidence is unquestionably in favour of the existence in the Ferns of distinct sexes, and of a process of impregnation which gives rise to a new individual. It has also been ascertained that something like what in the animal kingdom is called an alternation of generations, takes place in the Ferns, the one complete generation consisting of the prothallus, which is developed from the spore, and bears the antheridia and pistillidia, through which fertilisation takes place; the other generation, resulting from this act of fertilisation, being totally different, much more developed, and producing stems, fronds, and spores. The facts from which these inferences are drawn have been variously stated by different observers. The development of the spores, according to Prof. Henfrey,* who has thoroughly investigated the subject, consists of a bursting of the tough outer coat of the spore, and a protrusion of the delicate inner membrane in the form of a little tubular pouch, the contents of which soon begin to acquire a green colour. This tubular process elongates, and at length becomes divided by cross partitions, a row of five or six cells being sometimes formed. The first formation of root fibres, which often occurs in the earliest stages, consists of the growth outwards of the wall of one or more of the cells of the filamentous portion of the prothallus into a long slender undivided tube, and the roots of the full grown prothallus exhibit the same characters, being tubular prolongations of the inferior walls of the cells of the green frondose expansion. After a time the youngest cell of the growing prothallus becomes more expanded in the transverse diameter, and after the next transverse division of the cavity, a new mode of increase occurs; the newest cell becomes divided in a direction parallel to the original direction of growth. It is by the repetition of these two modes of extension that the prothallus gradually acquires its flat expanded form. When the prothallus has attained some size, the cells about the middle of the front border are produced smaller than those at the sides and anterior angles, so that the latter advance forwards as rounded lobes, leaving a sinus, or notch, in the centre, the prothallus thus acquiring a kind of inversely heart-shaped figure. Numerous radical hairs are produced about the posterior part of the prothallus, and in the same part a process of horizontal cell division is commenced, so that a thickness of two, three, or more cells is formed.

If a leafy stem is produced, the prothallus dies away; but if none of the pistillidia become fertilised, so that the prothallus remains barren, its vegetative existence may be indefinitely prolonged. In this case the lateral lobes enlarge, and their margins become variously curved, sinuous, or convoluted; and new lobes sometimes

* Henfrey, *Transactions of the Linnæan Society of London*, xxi. 117.

form from the cushion-like thickening of the posterior part, or individual cells of the margin grow out, and repeat the mode of development exhibited by the spore-cell in the original germination, producing new prothalli by a process of budding. These "proliferous" prothalli bear antheridia, but the pistillidia have not been observed upon them.

At an early but variable period of the growth of the prothallus, a number of small cellular bodies, analogous to the pollen of flowering plants, are produced on its lower surface, and chiefly about its central thickened part. These, which are called *antheridia*, consist of globular cells enveloped by one or two annular cells, and containing a number of free cellules or vesicles or sperm-cells, within which a spiral fibre representing the *spermatozoid*, is coiled up. These spermatozoids consist of a flattened band, curled spirally in about three and a-half coils, and bearing along the outer edge cilia of considerable length. If these parts are placed in water, as they must be for the purpose of observation, the cilia are seen to vibrate with such rapidity, that when the spermatozoid is in active motion they appear only as a fringe of light. The motion of these bodies consists of a rapid rotation around their axes, and this combined with their spiral form induces a forward movement of great velocity; the movement is, however, without regularity, the spermatozoids darting here and there, or turning aside without rule. If they come in contact by the smaller end, with any fixed body, they often adhere by this point, and then revolve around their axes without advancing. By degrees the motion slackens, and the rotation is lost, a kind of vibratory motion only remaining, and this at length ceases; but the spermatozoids appear to undergo dissolution during this time, and when they come to rest often appear as shapeless masses. When the motion is artificially arrested, which is done instantaneously by applying iodine, the flattened band is seen to have a little rounded head, from which the coil runs back, increasing successively in diameter, so as to give the whole a cone-shaped outline in the side view.

It will thus be seen that, according to Professor Henfrey's explanation, the antheridia consist of a large cell, having thick walls consisting of a double membrane, and containing free cellules, which are the spermatozoids coiled up within a membrane, from which he has found them in some instances apparently unable fully to extricate themselves. We learn, however, from Dr. Lawson, that his observations lead him to conclude that it is the cell-wall of the free cellules, which itself unrolls to form the spiral filaments, and that they are not invested by any external membrane.

At a later period of the growth of the prothallus, other larger and more complex cellular bodies appear, which are analogous to the ovules or nascent seeds of flowering plants. These bodies, called *pistillidia*, or *archegonia*, are much fewer in number than the antheridia. They are formed near the centre of the prothallus, and at first consist of a single cell of the (inferior) surface, destined to become an *embryo-sac*. This, by a process of cell division, becomes enlarged, so as when full grown to acquire the appearance of papillæ, of a bluntly conico-cylindrical form, and composed of about four tiers of four quadrant-shaped cells. In the embryo-sac at the base, a central globular body, which forms the *germinal vesicle*, is at length produced, and a clavate cavity or canal extends from this between the convergent inner angles of the four series of cells, up to the apex of the papillæ. The germinal vesicle becomes fertilised through the agency of one of the spermatozoids, or spiral filaments, which enters by the canal, and an embryo plant possessing a terminal bud is gradually developed.

With regard to the fertilisation, Professor Henfrey is of opinion that the operation is effected by the contact of one or more of the spermatozoids with a mucilaginous filament contained in or hanging from the mouth of the canal of the pistillidium. I have, he observes,* seen the spermatozoids swimming in numbers around the mouth of the archegonia, but never detected one inside, and I do not see any good reason for supposing such a process necessary. The pollen-tube of flowering plants only comes in contact with the outside of the embryo-sac, and the influence is sometimes communicated through a long suspensor. There does not therefore seem to be any sufficient objection to the supposition that the contact of the spermatozoid with the filament of mucilage which lies in the canal of the archegonium, suffices to convey the necessary stimulus.

Notwithstanding some differences in the statements of those who, by the aid of the microscope, have investigated this subject—differences of no importance in a practical point of view, and merely presenting questions of anatomical and morphological interest, the balance of the evidence they afford is strongly in favour of the view now generally taken, that in this family of plants there are organs analogous to the sexual parts of other plants, and that the united agency of these parts is necessary to the production of new individuals. The whole subject is very amply and lucidly treated in Professor Henfrey's memoir, from which most of the foregoing statements are derived, and which, with its accompanying illustrations, should be consulted by all who take interest in the question.

After the organisation of a terminal bud within the pistillidium, young fronds soon make their appearance. These, at first, are very unlike those of the mature plant, being of more simple form and more delicate texture, but they gradually acquire more and more the characteristics peculiar to their species, though, with the exception of a few annual kinds, they are a year or two, or, in many cases longer, in arriving at a perfect state.

* Henfrey, *Transactions of Linnæan Society of London*, xxi. 125.

The production among the Ferns, especially in the genus *Gymnogramma*, of certain intermediate forms apparently hybrids, is suggestive that something like hybridisation occurs among the cryptogamous plants, but this is a matter rather of inference than of certainty; for it may be that in these cases, the change of form and character which is seen to be produced, is rather the result of the development of a sporting bud, than a product of actual hybridisation; while the more frequent occurrence of these intermediate forms in certain genera, may result from the tendency to sport being more strongly developed in these than in others.

There is much opportunity for interesting observations in tracing the successive stages of the growth of the prothallus, and the gradual development therefrom of the young plant—at first a tiny green speck, then a thalloid scale, and finally a bud with its attendant fronds. Those who have leisure for microscopic research, without the patience or skill to undertake the more abstruse physiological inquiry involved in the development of these parts, will find it an occupation full of interest. Thus, to watch the tiny atoms as they spring into active life, and day by day acquire bulk and strength, must lead a thoughtful mind to pure and wholesome reflection; inasmuch as it must result in the conviction that life thus interminably renewable from the dust-like spore, can have but one source—that source the Author and Giver of all life.

CHAPTER IV.

CLASSIFICATION OF FERNS.

NOMENCLATURE AND ARRANGEMENT: In what sense important—Characters employed for generic distinctions—Venation as a generic character—Mr. Smith's proposed arrangement: Eremobrya and Desmobrya—Their peculiar structure—Mr. Newman's scheme—The principal groups of Ferns, and their characteristics—Classification proposed to be adopted.

THIS is a most important branch of our subject; for without something like a correct notion of the principles and the characters on which the classification of Ferns is founded, it is impossible to have more than a vague and uncertain knowledge of their nomenclature. The species of Ferns inhabiting Great Britain being few in number, they afford but little aid in illustrating the principles on which pteridologists depend in the naming and arrangement of the Fern family; so that to give an intelligible view of the subject, we must briefly sketch the classification of Ferns in general.

Though in some sense important, the Nomenclature or naming, whether scientific or popular, of natural objects, has, it is always to be remembered, no other value than that of enabling different observers to recognise the same objects, and to impart to each other information respecting them; and Arrangement or Classification has no other importance than that of facilitating these objects. Hence, it is as a means to an end, that we refer to classification as an important branch of our subject, that end being the identification of genera and species, the main object of such identification being the intercommunication of information concerning them. It is not for a moment to be supposed, that the mere naming and the classifying of the objects which engage the attention of scientific men, are the aim and purpose of study, as some seem to think, or to act as if they thought. These are, in fact, but the beginning, the very alphabet of inquiry, and are only useful in so far as they facilitate the acquisition of real knowledge respecting the structure, the habits, the beauties, the properties, the uses, &c., of those portions of the Creator's wonderful handiwork which may come under observation; in so far, moreover, as they may serve as indices to the scattered information which human research has drawn, as from a deep unfathomable well, concerning them. This, then, is the sense in which nomenclature and arrangement are to be understood as important.

The characters originally employed in the grouping of Ferns into groups resembling our modern genera, were derived from the shape and division of the fronds, but these were found to produce very unsatisfactory associations. The generic, or family characters, were next sought for in the organs of reproduction. When these came to be adopted, the shape of the sori, or clusters of spore-cases, was taken as the most obvious feature; but though affording better discriminative marks than the outline and division of the frond, the mere form of the sori still proved to be insufficient. The presence or absence of an indusium or cover to the sori, and the form and attachment of this cover when present, were next taken into account; and this combination of characters derived from the sori formed the principal advance made up to the early part of the present century. Something more was, however, found to be required, as the knowledge of species became more extended, and this was at length found in the peculiarities of structure presented by the venation of the frond, and in the connection of the veins with the sori.

The names of Langsdorf and Fischer, of Brown and of Brongniart, the latter in connection chiefly with fossil Ferns, stand pre-eminent amongst those who at an early period recognised in these structural peculiarities, features of generic importance. Others, as Bory, Gaudichaud, Kaulfuss, Blume, &c., made use to some extent of characters derived from the venation; but subsequently characters derived from this source have been extensively employed, especially by Presl, J. Smith, and Fée, and arrangements founded thereupon are generally adopted at the present day. Indeed, this feature is made the basis of most modern methods of classifying Ferns.

We have elsewhere* quoted the remark of the late Robert Brown, to the effect that for the purpose of subdivision, "the most obvious as well as the most advantageous source of character, seems to be the modifications of the vascular structure, or the various ramifications of the bundles of vessels or veins of the frond, combined with the relation of the sori to their trunks or branches."† That learned and lamented botanist seems indeed to have had chiefly in view the formation of sub-genera, but the whole of the series of groups known as genera, sub-genera, and even species, are so highly artificial and arbitrary, that the exact use to be made of these characters becomes a mere question of words and of convenience. At least in cases where the groups are so large and varied as to require the adoption of many sub-genera, as in the ancient *Polypodium* for example, it becomes less complex to consider the lesser divisions as genera, rather than as sub-genera, and the one course is as correct, technically viewed, as the other. To us, therefore, it seems most convenient to adopt, but not to excess, the general principle of employing the venation as a supplementary generic character.

Then, as to its value. Looking at the question of venation as illustrated by the great and universally adopted natural divisions of flowering plants, its use as a source of generic character in the case of the Ferns rests upon better ground than that of mere convenience. The constant and unvarying occurrence of parallel free

* Moore, in *Proceedings of the Linnæan Society*, ii. 210.

† R. Brown, in Horsfield's *Plantæ Javanicæ*, 3.

veins and of reticulated veins in the primary groups of flowering plants, and the value assigned to them, are significant facts; and importance may be equally claimed for similar differences among the Ferns. That some auxiliary character beyond that afforded by the sori and their covers, is required, is very commonly admitted; for while the complete series of floral organs in flowering plants offer such numerous and varied characteristics for generic combination, the differential characters to be drawn from the sori in Ferns are exceedingly limited. And what auxiliary character is there so proper to be employed as the vascular structure of the plant with which the sori are so intimately connected? Experience, moreover, attests that it may be relied on with perfect confidence, for it is found, with a few insignificant exceptions, that whatever condition of the venation occurs in a particular species, that condition is constant to that species.

The vascular system, as we have already stated,* must be regarded as of the highest importance in the economy of plants, even in reference to their propagation, for cases are not at all infrequent in which certain extraordinary means of development, namely, adventitious buds, are formed in direct connection with it. In the Ferns particularly, those points of the veins which normally serve as the receptacles to which the spore-cases are attached, in other cases become viviparous, and develope gemmæ or buds instead of spores.

Though thus claiming systematic importance for the venation in Ferns, and supporting its use as a source of generic character, we are ready to admit that the question is not altogether free from anomalies, or without its difficulties; these are, however, not greater than occur in the application of our imperfect knowledge to the classification of other groups of plants, where even with all the variety of character afforded by the flower and seed, anomalous and dubious species are not uncommon; nor are the anomalies here

* *Proceedings of Linnæan Society*, ii. 211; and *Gardeners' Chronicle*, 1853, 86.

more difficult to overcome, than those presented by all other methods which have been proposed for the classification of the Ferns.

We have already recorded* some instances in which the sori of dorsiferous and other Ferns were aberrantly situated. The inferences to be drawn from the examples referred to are, we think, confirmatory of the importance of the venation. It would thence appear that the veins are important structures in the economy of Fern development, since they are capable of originating the receptacle and spore-cases from their surface in any part—even in unusual parts—of the frond. This being so, sufficient importance would appear to attach to them, to justify their being employed for the purpose of assisting in the definition of genera, in a family of plants where something more than the so-called fructification itself is confessedly needed to supply distinctive characters.

Mr. J. Smith has recently proposed† to classify the Ferns according to their mode of development from the caudex or rhizome. Taking advantage of the apparent difference between the growth of those Ferns which have a caudiciform and a rhizomatous stem, he proposes to bring the polypodiaceous Ferns under two groups, which he calls *Eremobrya* and *Desmobrya*.

In the *Eremobryæ* "the fronds are developed from the sides of a special rhizome which has its axis of growth always in advance of the nascent frond (excurrent); the fronds are produced from nodes more or less distant from each other, each node producing a single frond, which after having arrived at maturity separates by a special articulation formed between the node and the base of the stipes. After the frond has fallen the node remains in the form of a round concave cicatrix generally more or less elevated. The rhizome is solid, fleshy, and brittle, varying from long and slender to more or less short and thick, and is always covered with scales which unless

* Ante, p. 22.

† J. Smith, *Botany of the Voyage of H.M.S. Herald*, 226.

they are common to the whole frond, seldom extend upwards beyond the node. This mode of development is peculiar to a considerable number of the *Polypodieæ*, including genera with both free and anastomosing veins; also a portion of *Davallia*, of which *D. canariensis* may be viewed as the type; the whole forming a truly natural group of Ferns."

In the *Desmobryæ*, "the fronds rise from a terminal axis, either in a single alternate series or in a fascicle forming a corona; each succeeding frond is produced on the interior side of the bases of the preceding fronds, the bases being united and adherent. By the successive evolution of fronds, a progressing accessory stem or cormus is formed, which varies in being decumbent or erect, short or more or less elongated, often assuming the aspect of trees, or creeping on or under the surface of the ground, frequently forming cæspitose tufts. In those species producing their fronds in a single series, the developing axis sometimes elongates before the evident evolution of the fronds, which are then more or less distant from each other; by this mode of growth is formed a creeping or scandent caudex which often assumes the character of a sarmentum, and then appears to agree with the mode of growth in the *Eremobryæ*, but it is readily distinguished by the epidermis and vascular structure of the stipes being continuous and united, forming part of the developing axis and not being articulated as in the *Eremobryæ*. The *Desmobryæ* include part of *Ctenopteris*, the whole of the *Phegopteris* group of *Polypodium*, also *Gymnogramma*, *Goniopteris*, *Meniscium*, and other genera of the *Polypodieæ*; the whole of *Pterideæ*, *Asplenieæ*, and with a few exceptions *Acrosticheæ*, *Aspidieæ*, *Dicksonieæ*, and *Cyatheæ*."

Such is Mr. Smith's definition of those groups, which he epitomises thus: *Desmobrya*—Fronds in vernation terminal, their bases adherent, united with and constituting the axis of growth: *Eremobrya*—Fronds in vernation lateral, solitary, and ultimately separating from the axis by a special articulation. The scheme, as hitherto explained, does not apply to the *Osmundaceæ*, *Ophioglossaceæ*, &c.

The adoption of this mode of classification would to a great extent break up the groups and genera now recognised; and we cannot think such radical changes, in the case of plants already so well classified as the Ferns, at all necessary or desirable. Probably the necessity which has been felt for such changes has arisen from a concentration of the attention on matters of detail, without pausing to form clear generalised perceptions of important differences. It must, moreover, be admitted in extenuation of proposed changes, that under the systems now in use, difficulties and objections here and there arise, such as a disturbance of the ordinary development of the veins, or the impossibility of determining on the presence or absence of indusia; but these are not more important than those which are constantly occurring in other departments of botanical science, nor more insurmountable than those which would be likely to occur in the application of any other set of characters by the light of our present limited and ever-varying knowledge. The same difficulties, too, would still occur at a subsequent stage of inquiry, even if other characters were resorted to for primary distinctions.

There is, moreover, no real physiological difference, as has been claimed, between the two apparently different modes of development which have been made the basis of this mode of classification. In both the axis is a stem, assuming in one case the form of a rhizome, in the other more or less that of a caudex or trunk—both being equally, forms of stem. Neither is the development of the fronds in the one case really terminal, though apparently so, for in the very nature of things, the axis must be developed before the part it supports. The original suggestion, at first sight, appears to produce a natural division, in some measure equivalent to that of Exogens and Endogens among flowering plants, but such a contrast is, in reality, inadmissible, the whole race of cryptogams going to make up the group of Acrogens, which is the real equivalent to the endogenous and exogenous groups of phænogams.

Mr. Newman, following Mr. Smith, has proposed a modification of his plan, which is, however, open to the same objections. He at first proposed to form four groups: *Eremobrya*—Ferns whose fronds are produced from any part of the rhizome except its point, and always articulated with it: *Chorismobrya*—Ferns whose fronds are produced as in the preceding, but not articulated: *Desmobrya*—Ferns whose fronds are produced only at the point of the erect or suberect corm-like rhizome, and not articulated: *Orthobrya*—Ferns having the vernation straight. This arrangement has been subsequently curtailed, and the following scheme substituted *—A primary group, the *Filicales*, including those Ferns which have the spore-cases encircled by a ring, are divided into two lesser groups:—the *Rhizophyllaceæ*, in which the fronds are attached to a rhizome, or root; and the *Cormophyllaceæ*, in which they are attached to a cormus or trunk. Another primary group, the *Osmundales*, including the rest of the Ferns, namely those which have their spore-cases detached from the leaves, and not encircled by a ring, comprise the *Osmundaceæ*, with circinate vernation, and a woody trunk; and the *Ophioglossaceæ*, with straight vernation, and a succulent trunk.

That system of classification which is based on the vascular system of the frond, taken in conjunction with its fructification, is, we think, in every respect to be preferred, and we shall here explain that modification thereof which we have adopted † after much consideration.

All Ferns, taking the term in its widest application, are referrible to one of these groups, namely—

OPHIOGLOSSACEÆ—POLYPODIACEÆ—MARATTIACEÆ.

Of these, the *Ophioglossaceæ* and *Marattiaceæ* are but small groups,

* Newman, *History of British Ferns*, 3 ed. x.

† Moore, *Index Filicum*, Synopsis ix.

while the *Polypodiaceæ* include the greater portion of all known Ferns. These three groups may each be regarded as a distinct order of plants, forming together the FILICES or Ferns.

The groups thus indicated are distinguished from each other by the nature and structure of their spore-cases. The presence of the *annulus* or ring around the spore-case, in some form or other, either nearly completely surrounding it, or in a more or less rudimentary condition, is the distinctive peculiarity of the *Polypodiaceæ*, while the *Marattiaceæ* and the *Ophioglossaceæ*, are distinguished by the absence of any such ring, rudimentary or otherwise. These obvious distinctions, to which, as they are connected with certain other differences which serve the same purpose, the initiated seldom have to resort, are the foundation of all satisfactory inquiries into the nomenclature of a Fern.

The *Marattiaceæ* and *Ophioglossaceæ* are distinguished from each other by very obvious characters. The *Marattiaceæ* are dorsiferous, that is, bearing their sori on the back or under surface of their fronds, as is commonly the case among Ferns. The *Ophioglossaceæ*, on the contrary, always have their spore-bearing or fertile fronds contracted, so that while the spore-cases are produced marginally, they seem to occupy the whole surface; such fronds are called *rachiform*, because they want the flat leafy expansion of ordinary fronds, and are reduced to the appearance of mere ribs or rachides.

The *Ophioglossaceæ* are few in number, and present little difference of structure, so that it has not been found necessary to range the genera in secondary groups or tribes. The *Marattiaceæ*, however, form three little groups of genera, separated by differences of some importance; these tribes are called the *Marattineæ*, the *Kaulfussineæ*, and the *Danæineæ*. The *Marattineæ* have their sori ranged in two lines facing each other, forming distinct oblong masses. The *Kaulfussineæ* have distinct circular sori, the spore-cases of each sorus being concrete into a single annular series, and furnished with openings towards the centre. The *Danæineæ* have their sori connate over the whole under surface of the fertile fronds, which then show long parallel lines of small round cavities, which are the openings of the concrete spore-cases.

The *Polypodiaceæ* offer so much variety of structure in their spore-cases that it becomes necessary to range the genera under eight tribes or divisions, distinguished chiefly by peculiarities in the form of the spore-cases, in their number and position, or in the structure and development of the annulus or ring, which latter presents some curious differences. These tribes are as follows:—

(1) *Polypodineæ*, the most extensive of all, in which the spore-cases are almost equally gibbous or convex on both sides, with a vertical and nearly complete ring, and bursting transversely at a part on the anterior side, called the stoma, where the striæ of the ring become dilated into elongate parallel cells.

(2) *Cyatheineæ*, in which the spore-cases are sessile or nearly so, and oblique-laterally compressed, the nearly complete ring being, in consequence, more or less obliquely vertical, that is, vertical below, curving laterally towards the top, bursting transversely, and seated on an elevated receptacle; they approach very near the *Polypodineæ* through some species of *Alsophila*, in which the characteristic obliquity of the ring is little apparent.

(3) *Matonineæ*, consisting of a single species only, in which the ring is broad, sub-oblique, and nearly complete, the spore-cases sessile, bursting horizontally, not vertically, the sori dorsal and oligocarpous, covered by umbonato-hemispherical indusia, which are peltate, that is, affixed by a central stalk; they may be compared to an inverted cup.

(4) *Gleicheninceæ*, in which the ring is complete and transverse, either truly or obliquely horizontal, the spore-cases being globose-pyriform, forming oligocarpous sori, *i. e.*, sori consisting of but few spore-cases (2-4 to 10-12), situated at the back of the frond, sessile

or nearly so, and bursting vertically; while the fronds are rigid and opaque, and are usually dichotomously-branched.

(5) *Trichomanineæ*, in which the ring resembles that of the *Gleichenineæ*, but the spore-cases are lenticular, numerous, clustered on an exserted receptacle, which is, in fact, a prolongation of the vein beyond the ordinary margin of the frond, so that the sori become extrorse-marginal, or projected outwards as well as opening outwardly; while the fronds are pellucid-membranaceous.

(6) *Schizæineæ*, in which the ring is either horizontal or transverse, but situated quite at the apex of the oval spore-case, which is, in consequence, said to be radiate-striate at the apex; the spore-case is also sometimes resupinate, or turned upside down, so that the true apex is below.

(7) *Ceratopteridineæ*, consisting of one or two, perhaps only a single aquatic species, in which the spore-cases are sometimes furnished with a very rudimentary ring, reduced, as in *Osmundineæ*, to a few parallel striæ, sometimes furnished with a very broad and more lengthened ring. In this little group, the spores themselves furnish an excellent supplementary characteristic, being bluntly triangular, marked with three series of concentric lines.

(8) *Osmundineæ*, which is distinguished from the rest by having its spore-cases two-valved, bursting vertically at the apex; the ring, moreover, is very rudimentary indeed, being reduced to a few parallel vertical striæ (parallel elongated cells of the tissue) on one side near the apex of the spore-case. In all the preceding tribes, the spore-cases are not valvate, and consequently, when they open for the liberation of the spores, they burst partially or irregularly, and do not split at the top in two equal divisions, as occurs in the *Osmundineæ*.

These primary and secondary groups, consisting of orders and tribes, will be more readily comparable in the following summary:—

Character	Group
Spore-cases ringless—	
Fructifications marginal on rachiform fronds	OPHIOGLOSSACEÆ.
Fructifications dorsal on flat leafy fronds	MARATTIACEÆ.
Sori oblong, distinct, longitudinally biserial	MARATTINEÆ.
Sori circular, distinct; spore-cases annularly concrete in a single series	KAULFUSSINEÆ.
Sori connate throughout the under surface of fertile fronds	DANÆINEÆ.
Spore-cases having a jointed ring	POLYPODIACEÆ.
Spore-cases not valvate—	
Ring vertical, nearly complete; spore-cases usually stalked, gibbous; bursting transversely; (receptacles superficial or immersed)	POLYPODINEÆ.
Ring obliquely vertical, nearly complete, narrow; spore-cases crowded, sessile or subsessile, obliquelaterally-compressed, bursting transversely; (receptacles elevated)	CYATHEINEÆ.
Ring sub-oblique, nearly complete, broad; spore-cases few, sessile, gibbous, bursting transversely; (sori oligocarpous)	MATONINEÆ.
Ring horizontally or obliquely transverse, complete; spore-cases sessile or subsessile, bursting vertically—	
Ring zonal; spore-cases vertically compressed—	
Sori dorsal; (fronds rigid)	GLEICHENINEÆ.
Sori extrorse-marginal; (fronds pellucid)	TRICHOMANINEÆ.
Ring apical; spore-cases oval, crowned by the convergent striæ of the ring, *i. e.*, radiate-striate at the apex, sometimes resupinate	SCHIZÆINEÆ.
Ring rudimentary or more or less incomplete, very broad, flat, obliquely-vertical; spore-cases sessile, globose	CERATOPTERIDINEÆ.
Spore-cases two-valved; (ring rudimentary transverse)	OSMUNDINEÆ.

Of the British Ferns three species only belong to *Ophioglossaceæ*, and all the rest to the *Polypodiaceæ*, those belonging to the latter order being referred—one to *Osmundineæ*, three to *Trichomanineæ*, and the remainder to the *Polypodineæ*.

The group *Polypodineæ* is divided into twenty-three sections, of which the following only are represented in the British Flora:—

Lomarieæ or *Blechneæ*, having linear indusiate intramarginal sori transverse to the veins.
Adianteæ, having oblong sori, borne on the under side of the indusium itself, and therefore resupinate.
Pterideæ, having linear indusiate marginal sori, transverse to the veins.
Asplenieæ, having oblong or linear indusiate sori, lying obliquely, parallel with the veins.
Gymnogrammeæ, having oblong or linear oblique naked sori.
Polypodieæ, having punctiform naked dorsal sori.
Aspidieæ, having punctiform indusiate dorsal sori.
Cystopterideæ, having punctiform covered dorsal sori, the indusium being semi-involucriform.
Peranemeæ or *Woodsieæ*, having punctiform involucrate dorsal sori.

These sectional groups, are arranged in the foregoing order, in our *Index Filicum*, the intention being to give precedence to those in which the fertile principle is most evident, these being considered as having their organs most perfectly developed.

The principal modern arrangements of Ferns which the student will derive advantage from consulting, are those of Professor Fée,* Mr. John Smith,† and the late Dr. Presl.‡ Of these, that of Professor Fée is both the most recent, and the most perfectly and carefully digested, but it applies only to the two groups we have above distinguished as the *Polypodineæ* and the *Cyatheineæ*, so that several of the smaller groups are not included. Another elaborate arrangement of still more recent date, is that of Professor Mettenius,§ who does not, however, adopt the modern genera to any extent, and whose grouping appears somewhat complicated.

* Fée, *Genera Filicum*, Strasbourg, 1850—1852.
† J. Smith, in Hooker's *Journal of Botany*, iv. 38 *et seq.*; *London Journal of Botany*, i. 419 *et seq.*
‡ Presl, *Tentamen Pteridographiæ*, Prague, 1836; *Hymenophyllaceæ*, Prague, 1843; *Supplementum Tentaminis Pteridographiæ*, Prague, 1845; *Epimeliæ Botanicæ*, Prague, 1849.
§ Mettenius, *Filices Horti Botanici Lipsiensis*, Leipsig, 1856.

CHAPTER V.

SYNOPTICAL TABLE OF GENERA.

The few genera which are represented among the British Ferns, afford so inadequate an illustration of any particular system of classification, that their sequence becomes altogether unimportant. In our *Index Filicum* we have attempted an arrangement of the entire family of Ferns, on the plan which has been explained in the foregoing pages; and this arrangement we follow here in its leading features, that is, in so far as respects the principal groups; but as regards the sectional subdivisions of these groups, which give the actual sequence of genera, it will be more convenient here to follow the order adopted in other works on British Ferns, especially in our *Handbook of British Ferns,* and in the Folio edition of the present work. The sectional groups are the same as those of the *Index Filicum*, their relative position only having been changed.

Natural Order **POLYPODIACEÆ**, or True Ferns.

The Ferns belonging to this Order have their fronds circinate in vernation; their fructification dorsal or marginal; and their spore-cases furnished with an elastic jointed ring.

* SPORE-CASES NOT VALVATE.

Tribe I.—**Polypodineæ.**—*Fructification dorsal; spore-cases without valves, bursting irregularly and transversely; ring vertical, nearly complete.*

§ i.—Polypodieæ.—*Sori round, naked,* i. e. *without proper indusia or scale-like covers, dorsal on the veins.*

1. **Polypodium.**—Sori punctiform, round or roundish, exposed on the plane under surface of the frond.

2. Allosorus.—Sori round or sub-oblong, becoming laterally confluent, hidden beneath the scarcely attenuated reflexed margins of the frond.

§ ii.—Gymnogrammeæ.—*Sori linear, parallel with the veins, naked,* i. e. *without indusia, dorsal on the veins.*

3. Gymnogramma.—Sori linear, forked below; (an annual or biennial Fern).

*** Ceterach which has linear sori, also has the indusium obsolete, so as to simulate naked linear masses. See this genus under § iv. Aspleniеæ.

§ iii.—Aspidieæ.—*Sori punctiform, rotundate, covered while young by scale-like indusia of the same form, dorsal on the veins.*

4. Polystichum.—Sori covered by circular peltate indusia, attached at their centre.

5. Lastrea.—Sori covered by reniform indusia, attached at the notch on their indented posterior side.

§ iv.—Aspleniеæ.—*Sori oblong or linear, covered while young by scale-like indusia of the same form, lateral on the veins.*

6. Athyrium.—Sori oblong, lunately curved, sometimes (especially the basal ones) hippocrepiform or horse-shoe-shaped, the indusia attached along their concave edge, the free margin fringed; venules free.

7. Asplenium.—Sori simple, linear or oblong, oblique; venules free.

8. Scolopendrium.—Sori double, *i.e.* in proximate oblique parallel pairs face to face, covered by elongate straight indusia, which open along the centre of the double or twin sorus; venules free.

9. Ceterach.—Sori simple, oblong, scattered, growing from the anterior side of the veins (except the lowest on each pinna which is on the posterior side), hidden among imbricated chaffy scales; indusium obsolete; venules reticulated.

§ v.—LOMARIEÆ or BLECHNEÆ.—*Sori linear, transverse to the veins,* i. e. *ranged longitudinally between the midrib and margins, covered by special linear indusia.*

10. Blechnum.—Sori forming a continuous line on each side of, and parallel with the midrib, and within the margin.

§ vi.—PTERIDEÆ.—*Sori linear, continuous, transverse to the veins, covered by the reflexed margin of the frond, which is altered in texture, becoming indusiate.*

11. Pteris.—Sori forming a continuous marginal line, covered by the indusium-like reflexed edge of the segments of the frond.

§ vii.—ADIANTEÆ.—*Sori oblong, rarely elongated, borne on the veins of the under surface of the transverse indusia, and laterally confluent; indusium formed of the reflexed margin of the frond, altered in texture.*

12. Adiantum.—Sori transverse, growing on the reflexed indusium-like apices of the lobes; hence resupinate.

§ viii.—CYSTOPTERIDEÆ.—*Sori punctiform, covered while young by cucullate indusia, affixed posteriorly, and inflected hood-like over them.*

13. Cystopteris.—Sori covered by ovate cucullate or hooded indusia, which are attached by their broad base.

§ ix.—PERANEMEÆ or WOODSIEÆ.—*Sori punctiform, involucrate,* i. e. *with the scale-like membrane fixed beneath the sorus.*

14. Woodsia.—Sori within involucres whose margin is divided into incurved capillary segments.

TRIBE II.—**Trichomanineæ.**—*Fructifications extrorse-marginal; spore-cases without valves bursting irregularly, clustered on veins (receptacles) which project from the frond, and are surrounded by urn-shaped or two-valved involucres; ring horizontal or oblique, complete.*

15. Trichomanes.—Receptacles exserted, sporangiferous at the base, where they are surrounded by urn-shaped involucres of the same texture as the frond.

16. **Hymenophyllum.**—Receptacles included within two-valved involucres of the same texture as the frond.

** SPORE-CASES TWO-VALVED.

TRIBE III.—**Osmundineæ.**—*Fructification marginal paniculate; spore-cases two-valved, opening at top; ring rudimentary near the apex, consisting of a few parallel striæ.*

17. **Osmunda.**—Fructification forming irregular, densely-branched panicles on contracted rachiform segments, at the apex of the fronds.

NATURAL ORDER **OPHIOGLOSSACEÆ**, OR ADDER'S TONGUE FERNS.

The Ferns belonging to this Order have their fronds plicate or simply folded in vernation; their fructification being paniculate or spicate; and their two-valved spore-cases without an elastic ring.

* FRUCTIFICATION PANICULATE.

18. **Botrychium.**—Spore-cases arranged in irregularly-branched panicles terminating a separate rachiform branch of frond.

** FRUCTIFICATION SPICATE.

19. **Ophioglossum.**—Spore-cases forming two-ranked simple spikes terminating a separate rachiform branch of frond.

THE feathery Fern! the feathery Fern!
It groweth wild, and it groweth free,
By the rippling brook, and the wimpling burn,
And the tall and stately forest tree;
Where the merle and the mavis sweetly sing,
And the blue jay makes the woods to ring,
And the pheasant flies on whirring wing,
Beneath a verdurous canopy.

The feathery Fern! the feathery Fern!
An emerald sea, it waveth wide,
And seems to flash, and gleam, and burn,
Like the ceaseless flow of a golden tide:
On bushy slope, or in leafy glade,
Amid the twilight depth of shade,
By interlacing branches made,
And trunks with lichens glorified.

ANON.

THE BRITISH FERNS.

GENUS I: POLYPODIUM, *Linnæus.*

GEN. CHAR.—**Sori** non-indusiate, circular or ovoid, superficial or immersed; the **receptacles** terminal or medial on the free veins. **Veins** simple or forked, from a central costa, or simple costæform in the ultimate segments; **venules** free.

Fronds coriaceous herbaceous or membranaceous, simple pinnatifid pinnate or bi-tri-pinnate, articulated or continuous with the rhizome, the pinnæ sometimes articulated with the rachis.

Stem rhizomatous or caudiciform—**rhizome** creeping; **caudex** short erect, or decumbent.

The British species of *Polypodium* belong to two distinct sections of the genus, of which the type is *Polypodium vulgare.* The distinctive features of the group consist in the presence of circular or punctiform sori, and in the absence of covers or indusia; and by these marks it is easily distinguished from other British Ferns: but when exotic species are taken into account, another peculiarity, that of the free or disunited veins becomes necessary to distinguish it from the various generic groups subdivided from it, which have the veins more or less, and in various ways, reticulated.

Even the two groups into which the British species naturally fall, have by some modern writers been regarded as distinct genera, the habit of the plants being chiefly relied on to furnish distinguishing characters. There seems to us, however, to be such a complete conformity in the character of the sori, and such a conformity also in the nature of the venation, so far as words can express its chief peculiarity—namely, that the veins branch out from a central rib, and are free or disunited at their apices, that we cannot adopt the views of those who would separate them.

The fructification of *Polypodium*, it has been already stated, consists of round or dot-like masses of spore-cases. In some species these are developed at the tips of the veins, the receptacle or point of attachment being either small and punctiform, or dilated into an obovate form. The terminal-fruited British species has the veins rather divergently forked, and those which are not fertile are tipped by a club-shaped or thickened point, which is often visible on the upper surface, forming a line of slightly sunken dots which are occasionally white with a kind of cretaceous exudation. In other species the receptacles are produced on the back of the veins, at a greater or less distance from their point, these forming the medial-fruited group. The veins in the latter group do not terminate in club-shaped apices, and the sori though usually round and dot-like, are occasionally somewhat elongated.

Mr. Newman, adopting what he calls the rhizophyllaceous and the cormophyllaceous groups of annulate ferns, which mode of grouping he admits to be "a division which literally halves" such genera as *Polypodium* and others, carries his notion to an extreme point by discarding altogether the name of *Polypodium*, and imposing three new ones. Thus he proposes *Ctenopteris* as the name of the common *Polypody*, the old name of which, in any case, should be held sacred; whilst *Gymnocarpium* is provided for *P. Phegopteris* and its allies; and *Pseudathyrium* for the plants represented by *P. alpestre*. *Ctenopteris* had already been suggested as a sectional group of *Polypodium*, and to this there can be no reasonable objection; and the group indicated by *Gymnocarpium* had been also proposed as another section under the name of *Phegopteris*. This latter name is adopted for the plants referred to the *Gymnocarpium* and *Pseudathyrium* of Newman, by the majority of those pteridologists who admit the generic importance of the group.

The genus *Polypodium*, with the limits already indicated, is a large group scattered over the whole world, and containing numerous species which are separable into about half-a-dozen sections, two only of which are represented among British Ferns. These two are:—§ *Ctenopteris*, and § *Phegopteris*, the former with terminal sori, and articulated fronds, represented by *Polypodium vulgare;* the latter with medial sori, and fronds adherent to or continuous with the caudex, represented by *Polypodium Phegopteris.* The other groups are:—§ *Arthropteris*, with terminal sori, and articulated fronds and pinnæ, represented by the *Polypodium tenellum* of New Zealand; § *Adenophorus*, with terminal sori on dilated receptacles, borne on costæform *i.e.* simple central veins, and adherent fronds, represented by the *Polypodium hymenophylloides* of the Sandwich Islands; § *Prosechium*, with terminal sori, and adherent fronds, represented by the West Indian and South American *Polypodium pendulum;* and § *Themelium*, with basal sori on the costæform veins, and adherent fronds, as in the *Polypodium tenuisectum* of Java and Peru.

The name of the genus is derived from the Greek *polys*, many, and *pous, podos*, a foot.

SYNOPSIS OF THE SPECIES.

§ **Ctenopteris.**—*Sori terminal punctiform; fronds articulated with the rhizome.*

1. **P. vulgare:** fronds oblong or ovate, deeply pinnatifid.
 — **semilacerum:** lower segments pinnatifid, upper ones serrated, fertile.
 — **omnilacerum:** segments all pinnatifid, not narrowed below, the lobules pyramidal or acuminate serrate, sparingly fertile.
 — **cambricum:** segments all pinnatifid, narrowed below, the lobules serrated, crowded, barren.
 — **cristatum:** fronds and segments multifid-crisped at their apices.

§ **Phegopteris.**—*Sori medial, punctiform or sub-elongated; fronds adherent to the caudex.*

* *Fronds pinnate.*

2. **P. Phegopteris;** fronds ovate-triangular acuminate, pinnate below; pinnæ pinnatifid, the lower pair deflexed, the upper ones confluent.

** *Fronds bipinnate.*

3. **P. alpestre;** fronds lanceolate; pinnules toothed, or pinnatifid; stipes short.
 — **flexile:** fronds narrow, lanceolate, flaccid; pinnæ short deflexed.

*** *Fronds ternate or subternate.*

4. **P. Dryopteris;** fronds pentangular-deltoid, smooth; stipes glabrous; rachis deflexed.
5. **P. Robertianum;** fronds erect, subternate, elongately pentangular-deltoid, glandular-mealy; stipes glandular.

THE COMMON POLYPODY.

POLYPODIUM VULGARE.

P. fronds linear-oblong, ovate-oblong, or ovate, acuminate, deeply pinnatifid, almost pinnate below, thickish, smooth; the lobes linear-oblong, bluntish or acute, obscurely serrate. [Plate I.]

POLYPODIUM VULGARE, *Linnæus, Sp. Plant.* 1544. *Bolton, Fil. Brit.* 32, t. 18. *Smith, Eng. Bot.* xvi. t. 1149; *Id., Fl. Brit.* 1113; *Id., Eng. Fl.* 2 ed. iv. 267. *Hudson, Fl. Ang.* 455. *Curtis, Fl. Lond.* ii. t. 5. *Deakin, Florigr. Brit.* iv. 37, fig. 1579. *Hooker & Arnott, Brit. Fl.* 7 ed. 581. *Babington, Man. Brit. Bot.* 4 ed. 419. *Sowerby, Ferns of Gt. Brit.* 9, t. 1. *Moore, Handb. Brit. Ferns,* 3 ed. 49; *Id., Ferns of Gt. Brit. and Ireland, Nature Printed,* t. 1. *Mackay, Fl. Hib.* 337. *Hooker, Gen. Fil.* t. 69 B. *Bentham, Handb. Brit. Fl.* 625. *Lowe, Nat. Hist. Ferns,* i. t. 38. *Schkuhr, Krypt. Gew.* 12, t. 11. *Willdenow, Sp. Plant.* v. 172. *Sprengel, Syst. Veg.* iv. 52. *Svensk Bot.* t. 37. *Flora Danica,* t. 1060. *Fries, Sum. Veg.* 82. *Ledebour, Fl. Ross.* iv. 508. *Koch, Synops.* 2 ed. 974. *Gray, Man. Bot. North. U. States,* 590, t. 9. *Presl, Tent. Pterid.* 179, t. 7, fig. 3. *Fée, Gen. Fil.* 235. *Nyman, Syll. Fl. Europ.* 431.

POLYPODIUM VITERBIENSE, *Boccone, Mus. di Piante,* 60.

POLYPODIUM BOREALE, *Salisbury, Prod.* 403.

POLYPODIUM OFFICINALE, *Guldenstadt, Reis. dur. Russ.* i. 421; ii. 25, 166.

POLYPODIUM PINNATIFIDUM, *Gilibert, Excerc. Phytolog.* ii. 577.

POLYPODIUM CANARIENSE, *Willdenow Hb.* 19647; *and of gardens.* (*A variety.*)

POLYPODIUM AUSTRALE, *Fée, Gen. Fil.* 236, t. 20 A, fig. 2. (*A variety.*)

POLYPODIUM VIRGINIANUM, *of gardens;* ? of Linnæus, who quotes under this name figures of Morison and of Plumier,—the former apparently representing a small state of *Polypodium vulgare,* the latter *Polypodium incanum,* —but has left no specimen in his Herbarium. (*A variety.*)

POLYPODIUM INTERMEDIUM, *Hooker & Arnott, Bot. Beech. Voy.* 405. (*A variety.*)

CTENOPTERIS VULGARIS, *Newman, Phytol.* ii. 274; *Id., Phytol.* 1851, *App.* xxix.; *Id., Hist. Brit. Ferns,* 3 ed., 41.

Var. **cristatum**; fronds and lobes multifid-crisped at their apices. [Plate IV.]

POLYPODIUM VULGARE, *v.* CRISTATUM, *Moore, Sim's Cat. Ferns,* 1859.

Var. **semilacerum**; fronds pinnatifid and fertile above, bipinnatifid below; pinnatifid lobes narrowed below; lobules distinct, linear, acute, serrate. [Plate V.]

POLYPODIUM VULGARE *v.* SEMILACERUM, *Link, Fil. Sp. Berol.* 127. *Moore, Handb. Brit. Ferns,* 3 ed. 50; *Id., Ferns of Gt. Brit. and Ireland, Nature-Printed,* t. 2 A.

POLYPODIUM VULGARE *v.* HIBERNICUM, *Moore, Handbook of Brit. Ferns,* 2 ed. 44. *Sowerby, Ferns of Gt. Brit.* 10.

POLYPODIUM VULGARE *v.* SINUATUM, *Francis, Brit. Ferns,* 4 ed. 22 (not of Willd.)

POLYPODIUM VULGARE *v.* SERRATUM, *Herb. Mus. Brit.*
POLYPODIUM VULGARE *v.* CAMBRICUM, *Smith, Eng. Fl.*, 2 ed., iv. 268 (in part). *Mettenius, Fil. Hort. Lips.* 31 (excl. syn.)

Var. **cambricum**; fronds barren, bipinnatifid throughout; lobes narrowed below, broader and pinnatifid in the middle; lobules crowded, linear, or linear-lanceolate, acuminate, serrate. [Plate VI.]

POLYPODIUM VULGARE *v.* CAMBRICUM, *Willdenow, Sp. Plant.* v. 173. *Bolton, Fil. Brit.* t. 2, f. 5 *a.* *Smith, Eng. Fl.* 2 ed. iv. 268 (in part). *Moore, Handb. Brit. Ferns*, 3 ed. 50; *Id., Ferns of Gt. Brit. and Ireland, Nature-Printed*, t. 3 A.
POLYPODIUM CAMBRICUM, *Linnæus, Sp. Plant.* 1546.
POLYPODIUM LACINIATUM, *Lamarck, Fl. Fran.* i. 14.
POLYPODIUM CAMBRICUM, *β.* CRISPUM, *Desvaux, Berlin Magazine*, v. 315; *Id., Ann. Soc. Linn. de Paris*, vi. 233.

Var. **omnilacerum**; fronds bipinnatifid throughout, sparingly fertile; lobes not narrowed below, but pinnatifid throughout, the lobules distinct, pyramidal, serrate. [Plate VII.]

POLYPODIUM VULGARE, *v.* OMNILACERUM, *Moore, Handb. Brit. Ferns*, 3 ed., 55; *Id., Pop. Hist. Brit. Ferns*, 2 ed. 66, 337.

Rhizome creeping, tortuous, branched, as thick as a swan's quill or one's little finger, densely clothed with ferruginous scales on a deciduous cuticle, the fibrous roots produced chiefly from the under side. *Scales* lanceolate or ovate, very much acuminated, crowded, sometimes peltately attached, at length deciduous, leaving the surface of the rhizome smooth and greenish. *Fibres* brown, tomentose, densely matted over the surface to which the rhizome is fixed.

Vernation circinate.

Stipes naked, variable in length, often nearly or quite as long as the frond, sometimes much shorter, and as well as the rachis slightly grooved in front; at the base articulated with the rhizome.

Fronds from two to eighteen inches long, lateral to the rhizome, subcoriaceous, of a somewhat sombre green, paler beneath; often triangular-ovate in outline when small, varying to ovate-oblong and linear-oblong, the latter being the form assumed by the fully developed condition of the species in its normal state; very deeply pinnatifid, usually more or less drooping. *Lobes* or *segments* linear-oblong, parallel, flat, blunt bluntish or abruptly acute, obscurely serrate, more distant and sometimes deflexed rarely shorter at the base, shorter and more crowded or becoming confluent near the apex, which sometimes terminates abruptly, but is usually caudate.

Venation in each lobe consisting of a prominent tortuous midvein or costa, which is alternately branched; the branches (*veins*) are again branched, producing from three to five alternate branchlets (*venules*). Of these venules, the lowest anterior one of each fascicle (rarely more) bears a sorus at its club-shaped apex, the others being sterile, and each terminating within the margin in a small transparent club-shaped head.

Fructification on the back of the frond, usually confined to its upper part, the sorus originating at the apex of the veinlet; at first a naked depressed scarcely visible spot, and from the earliest period at which it becomes visible quite destitute of any apparent membranous cover, or indusium. *Sori* or *clusters of spore-cases* circular, rarely somewhat oblong, quite exposed, arranged in a linear series on each side the midvein; at first distinct, often crowded and finally confluent. *Spore-cases* yellow or orange of various shades, becoming tawny, numerous, globose, with a slender stalk of elongated cells. *Spores* yellow, muriculate or corrugate, oblong or kidney-shaped.

Duration. The rhizome is perennial. The fronds are produced about the end of May, and are persistent through the winter and until after new fronds are produced, so that the plant is evergreen unless the fronds are destroyed or damaged by severe frost. Other fronds are produced later in the summer.

This common plant is the type of the Linnæan genus *Polypodium;* and as there is no reason, other than the fancy of name-makers, why that genus should be abolished, though there may be reasons for its reduction by divesting it of ill-assorted species, we cannot concur with those who give to this plant the name *Ctenopteris,* used for sectional distinction by Blume and Presl, and who thus altogether ignore the *Polypodium* of Linnæus. Whatever additional names the introduction of modern systems of classification may render necessary, it is clearly not permissible that the names of the types of *bonâ fide* established genera, where these can be recognised, as in this case, should be wantonly changed; and it may be well to remind those who are easily led either to make or to adopt changes of this nature, that names are not the ultimate objects of botanical or other scientific investigations.

The Common Polypody differs essentially from all the other British species associated with it, in the character of having its fronds articulated with the rhizome—that is, attached in such a manner that they separate spontaneously as they approach decay. Its texture, too, is stouter and firmer than that of the other native species, and in its normal form, it is, moreover, less divided than they. The small specimens produced on walls, and in other dry exposed places, are erect and rigid in habit; but in situations where it grows with more vigour, the plant becomes drooping and picturesque in character, indeed when growing vigorously from an old pollard stump, or among the roots of hedge-row plants on shady banks, it assumes quite an ornamental aspect.

This species, rupestral and sylvestral in its predilections, is one of our commonest and most abundant species, growing on rocks, banks, old walls, and tree stumps throughout Great Britain and Ireland; in other words extending laterally from Cornwall, Hampshire, Sussex, and Kent, its southern limits, to Shetland, Orkney, and the Hebrides, its northern boundaries. It is moreover distributed vertically from the coast level in the west of England, to an elevation of about 2100 feet in the Highlands of Scotland.

This common English Fern appears to be also abundant over Europe, extending from the Scandinavian kingdoms throughout Central and Western Europe, to Italy: Sardinia, Sicily, and Corfu on the Mediterranean side, and to Spain and Portugal on the Atlantic side: whence it extends into Africa by the Azores, Madeira, and the Canary Isles, occurring along the northern shore of the Continent, as at Algiers, and again appearing in South Africa, in the country of the Kafirs. In Asia, it is found in Siberia, and thence it extends eastwards to Kamtchatka and Japan, and westwards over the mountains of Western Asia to Erzeroum, but there is no certain information of its occurrence either in China or India. In North-west America it is widely dispersed, being found at Port Mulgrave,

Sitka, and the Slavo River; and thence through Columbia to Canada and the United States. The *Polypodium intermedium*, of Hooker and Arnott, from California is larger, and has the sori somewhat oblong, and the veins occasionally, though very rarely, anastomosing near the margin; but as this occurs in so slight and inconstant a degree as if by accident, and the oblong form of sorus is met with in native forms of *P. vulgare*, we can scarcely believe it to be more than a variety of the common species. The same may be said of the *Polypodium Karwinskianum*, of A. Braun, a native of Mexico and Guatemala.

The antiquated medicinal reputation of the Polypody seems to have little foundation in fact. It is supposed to be the "rheum-purging Polypody" of Shakspere. In some country places a decoction of the fronds is still used in the case of children as an expectorant remedy for colds and hooping cough, the mature fertile fronds, under the name of Golden Locks and Golden Maidenhair, being gathered for this purpose in autumn, and hung up like other herbs to dry; these are, when required for use, slowly boiled up with coarse raw sugar until the decoction becomes slightly bitter. The rhizome, which has a kind of bitter-sweet taste, when freshly infused, is reputed as a mild laxative; and, according to Dr. Deakin,* the Italians use common Polypody, in the form of decoction, as a demulcent. M. Desfosses found the rhizomes to contain a sweet substance resembling sarcocollin, mannite, incrystallizable sugar, starch, albumen, malic acid, magnesia, and oxide of lime. Pliny states that the root dried and powdered, and snuffed up the nose, will consume a polypus; and according to Dr. Lindsay,† the dried powdered rhizome was formerly applied externally as an absorbent, and for covering pills.

The fronds of this Fern contain a considerable quantity of carbonate of potass, which is sometimes obtained for use by burning. The ashes of the plant are then boiled in water, and the liquor is strained, and evaporated until the crystals are formed.

This Fern is easily cultivated, but it requires that light porous soil should be used, and that the rhizomes should be kept on the

* *Florigraphia Britannica*, iv. 39. † *Phytologist*, iv. 1065.

surface of the soil. When planted unnaturally deep, or in stiff retentive soil, it dwindles, and often eventually perishes. The most suitable compost is formed of leaf-mould, peat, and sand. Referring to its natural choice of situation, Mr. Newman observes *:—"It leaves the forest tree to rejoice in its vigour, but surrounds with a verdant crown the pollard willows that fringe the margins of our mill-streams, or overshadow our horse-ponds." Less happily he continues—"It is emphatically a parasite, a parasite moreover on the weak, and when it occasionally makes its appearance far away from man, and the works of men's hands, it is sure to be found clinging to some giant of the forest that is hastening to ruin." This circumstance of its being frequently met with growing on pollard trees, does not, however, give the Common Polypody a parasitical character in the proper sense of that term; it merely proves it to be sometimes epiphytal in habit, and as the plant is often found, fully as vigorous, growing among porous earth and on sandstone, such conditions are probably all merely accidental, the essential ones being a constant supply of moisture more or less in quantity, perfect drainage, and moderate shade. The plant will even exist in health, naturally, with little or none of some of these conditions about it, as many an old wall bears evidence.

The epiphytal habitats of this species indicate a mode of culture which is found to be successful. Like epiphytal orchids, these epiphytal Ferns are found to grow well suspended in open shallow baskets, the roots being protected by means of sphagnum moss, very light peat earth, and silver sand. The baskets should be of hazel or ash rods, or of copper wire, and with wide interstices; their form should be broad and shallow, resembling a saucer. In planting, a layer of the moss should be laid at bottom, and on this some of the rhizomes adjusted so that their points and the young fronds may readily push outwards. On these should come a stratum of the soil and moss intermixed, and near the top another layer of rhizomes partially covered with sphagnum moss. The moss is to be packed firm and kept so, by fixing a few cross bars at top. The whole should then be well saturated by dipping it in soft water, and this may require to be repeated at intervals during

* Newman, *History of British Ferns*, 3 ed. 43.

summer, or the necessary supply of moisture must be applied by means of syringing. A cool shady Fern house is the proper place in which to carry out this mode of culture. The fronds push out from the basket on all sides, and have a very pleasing appearance. This species may be readily increased by dividing the rhizome.

There are many deviations from the typical form of this species besides those which have been already briefly noticed; but they are chiefly of interest to the horticultural enthusiast rather than to the botanist; except, indeed, so far as the latter may regard them as evidences of the mode and extent of variation to which common species are subject, and may hence learn to appreciate rightly the differences which are found to exist amongst less familiar exotic species. It is, therefore, chiefly for the information of Fern cultivators, most of whom take an especial interest in these variations, that they are here enumerated.

The typical form of the Common Polypody has longish and comparatively narrow fronds. That form of this typical series which differs in the least degree, albeit constantly, from the normal state, has the ends of its lobes gradually tapering off to a narrow point, instead of being equal in width nearly to the end, and there more or less blunt. Another modification has the points of the lobes acute, but the margins are at the same time deeply notched, the notches forming a series of coarse double serratures; in this state, which has sometimes a tendency to furcation at the tips of the lobes, the sori are not unusually decidedly oblong, a remarkable feature, in which respect it somewhat deviates from the generic type. Another slightly differing form has the ends of some or all of the lobes divided, with the parts spreading, so that the lobes become more or less manifestly two-forked; or occasionally more than two points are developed to each lobe, and we have thus an indication of the manner in which are formed the tasselled apices which are now found to be common among British Ferns, and even occur in the present species.

The fronds, in some forms, moreover, acquire unusual breadth, so as to assume a broad oblong or ovate-oblong outline; and this departure from the typical outline is occasionally accompanied by various degrees of marginal division in the primary lobes, showing a tendency towards the more highly developed bipinnatifid varieties represented by *semilacerum* and *cambricum.* The most simple condition of this abbreviated and widened form, in which the apices are usually acute and the margins finely serrated, is almost or quite identical with the North American plant called *P. virginianum,* and nearly so with the Madeira plant called *P. canariense* in gardens; it is when deeply crenato-lobate, that this type of variation, which also sometimes varies with oblong sori, approaches the more highly developed or compound forms above alluded to.

We have thus indicated two different types of development among the numerous variations of the Common Polypody. In one of these the narrow elongate outline of the normal form is more or less preserved; and in the other, a tendency to develope breadth rather than length, results in a frond of broad outline comparatively short. The constancy of the varieties, as tested by cultivation, varies considerably according to the mode of treatment, and depends much upon whether they are kept confined or exposed; it is also influenced by the circumstance of the plants being established or otherwise, recently disturbed plants often running out or reverting to the normal state, but again assuming the character of the variety as they get established at the root. The only kinds, perhaps, that are unvarying in their characters, under all circumstances, are *omnilacerum* and *cambricum.*

Typal or oblong-fronded Series.

1. *acutum* (M.). This variety has the lobes without serratures, and narrowed gradually to a longish taper point. In its most marked condition it is not a common plant, but is very elegant, the fronds being rather broader than the common form of the species. Other less marked forms are not so much tapered, and are, moreover, slightly serrulate, approaching towards the acute-lobed state of the normal form. We have this variety from two localities in the neighbourhood of Guildford, Surrey; and from Cornwell, in Oxfordshire; and Mr. Clapham reports it from Settle, Harrogate, and Helmsley, in Yorkshire. [Plate II A.—Folio ed. t. 1 E.]

2. *bifidum* (Franc.) This has the lobes generally bifid or two-

forked, but sometimes three-forked or multifid. This lobing is generally unequal and irregular, and occurs mostly on the lower third of the frond, sometimes reaching two-thirds upwards, and occasionally, but very rarely, nearly to the apex. In its best state this is really handsome, the lobes being almost symmetrically bifid. It is not unfrequent. We have received specimens from—Kent: Blackheath; Tunbridge Wells, *Mrs. Delves;* Shettlefield, *F. Brent.* Surrey: St. Martha's Hill, near Guildford. Sussex: Hastings. Devon: Lustleigh, *Rev. J. M. Chanter.* Yorkshire: Byland Abbey, *C. Monkman;* Hutton near Malton, *C. Monkman* (large sub-oblong sori). A rare and more fully developed form, becoming ramified in the rachis, has been called *ramosum,* but it is not constant. [Plate III A.—Folio ed. t. 1 F.]

A serrated form, which may be distinguished as *bifido-serratum,* is found at Ruthin, *T. Pritchard,* and Malton, *C. Monkman.*

3. *interruptum* (Woll.) This form has the lobes interrupted or irregular, some being often here and there entirely wanting or much abbreviated; others variously and irregularly bifid or multifid, lobed or laciniated. This variety is rare: we have received it from—Surrey: Albury, *S. F. Gray.* Devonshire, *Rev. J. M. Chanter.* Yorkshire: Castle Howard, *C. Monkman.* A fine analogous form has been found at Barnstaple, in Devonshire, by *Mr. C. Jackson.*

4. *sinuatum* (Willd.) This when well marked is a prettily varied form, and very distinct; the lobes are curiously irregular, sinuous or waved, sometimes divided at the apices, but throughout irregularly lobate serrate or serrulate on the margin, the lobules themselves when large being often serrate. It is a rare form, and under culture for some years, has proved permanent. The plants have been found at—Tunbridge Wells, *G. B. Wollaston.* Devonshire, *Rev. J. M. Chanter.* Hest Bank, near Lancaster, *C. Monkman* (margins sinuated but not serrulate). Stirling, *Mrs. Macleod.* The same variety received from Bingley, has been figured by Bolton, t. 2, fig. 5 b. [Plate III C.]

An analogous form, chiefly remarkable for having a few coarse teeth near the ends of the lobes (*apicidentatum*) has been found near Stirling, *Mrs. Macleod.*

5. *serrulatum* (Woll.). This variety has the teeth of the lobes minutely serrate, which is its chief peculiarity, so that it does not deviate widely from the normal states of the species, though it has proved constant under cultivation. This has been found in—Devonshire: near Ottery St. Mary, *G. B. Wollaston.* Yorkshire: near Malton, *C. Monkman.* Carmarthen, Wales.

6. *marginatum* (Woll.). This form of variation consists in the splitting, in a very irregular way, of the epidermis on the margins of the lobes, one portion, generally on the under side of the frond, receding as it were towards the midvein; the lobes themselves are irregularly lacerate or serrate. Kent, *G. B. Wollaston.* Windermere, *F. Clowes.*

7. *auritum* (Willd.). This variety has at the base of the lobes on their anterior side, that is next the rachis on the *upper* margin, a distinct lobule or auricle, forming a kind of ear, from which it takes its name. The auricle is variable in respect to its appearance and size. Windermere, *F. Clowes.* Another aurite form with more acute lobes, is found by Mr. Clapham at Settle. [Plate II B.]

8. *compositum* (M.). This, so far as yet known, is a small growing variety, and combines peculiarities belonging to several other kinds; the lobes are crowded, broad and overlapping at the base, deeply lobed, the lobules becoming shorter upwards so as to give the lobes a pyramidal or in some cases a somewhat hastate outline; the anterior basal lobule is much larger than the rest, as in *auritum;* the upper lobes are less deeply divided but sinuated, as in *sinuatum;* while the division of the lower lobes in some degree resembles that of *semilacerum,* only they are broadest at the base. It is a very distinct and peculiar form; and was found near Ilfracombe, by the Rev. J. M. Chanter. A similar plant, probably the same, has been found by Mr. Elworthy, near Nettlecombe.

9. *cristatum* (M.). This variety has the points of all the lobes multifid-crisped, and the apex of the frond itself more or less ramose, with the branches crisped and tasselled. It is an exceedingly pretty and distinct variety, hitherto constant in cultivation; and was received from a correspondent in Ireland, by Mr. Sim of Footscray. A much less marked form of the same character, but which has not proved constant, was found in the Isle of Man by Dr. Allchin. [Plate IV.]

10. *serratum* (Willd.). The lobes in the typical state of this form are sharply and deeply serrate or even biserrate along their margins, and the apices are acute. The fronds are nearly of the usual outline, and with a tendency in the sori to become oblong. It varies, however, with the fronds rather broader, and the teeth somewhat rounded and sometimes partially enlarged into lobes, connecting this form with the variety *crenatum*. It has been found in—Kent: Sidcup, *G. B. Wollaston;* Tunbridge Wells, *Mrs. Delves.* Surrey: Godalming, *H. Bull;* Woking; Mayford; Shere. Sussex: Balcombe, *S. O. Gray;* Hastings, *Dr. Allchin.* Somerset: Cheddar; Nettlecombe, *C. Elworthy.* Devon: Moreton Hampstead, *Rev. J. M. Chanter.* Cornwall: Saltash, *Rev. C. Trelawny.* Monmouth: Chepstow, *S. F. Gray.* Hereford: Whitchurch; Mordiford. Warwickshire: Moseley near Birmingham. Gloucestershire. Oxfordshire: Cornwell, *H. Buckley.* Worcestershire: Malvern. Yorkshire: Byland Abbey, *C. Monkman.* Pembrokeshire: Castle Malgwyn, *W. Hutchison.* Denbighshire: Ruthin, *T. Pritchard.* Kirkcudbrightshire. Stirling, *Mrs. Macleod.* Galway. Clare: Ballinahinch. Waterford: Blackwater. Guernsey, *C. Jackson.* [Plate III B.—Folio ed. t. II B.]

A finely serrulated biserrate plant (*biserratum*) referrible to this form of the species, occurs at Eltham, *G. B. Wollaston.*

Broad or ovate-fronded Series.

11. *denticulatum* (M.). This form has fronds less coriaceous than usual, of a broad oblong outline, abrupt from the uppermost lobes not diminishing but terminating suddenly below the caudate apex; all the lobes are ascending, and notched with rather distant small sharp teeth. It has been found near Hereford, *Dr. Allchin;* and on Rochester Castle, *C. A. Chanter.* [Plate III D.]

An analogous but still broader form, having an ovate outline, found in Portugal, the Canary Islands, &c., is called *canariense,* and fronds resembling this have been gathered in Ireland, *e. g.* Lough Gill, Sligo, and Killarney, *R. Barrington.* Another similar form is the *P. virginianum* of North America.

12. *crenatum* (Woll.). This is a large growing form, approaching *semilacerum* by its broad or ovate fronds, and lobed segments; but

it varies considerably, in some instances approaching *serratum*, in others having the lobes more or less deeply and unequally crenate or crenato-lobate, and occasionally having the sori oblong. The notches of the lobes are however rounded, not acute, as in *serratum;* and the lobes themselves are in the typical forms prettily waved or undulated. The finest undulated form is a garden plant of obscure history. It is not common, but specimens we are disposed to refer to this variety, have been found in—Kent: Saltwood Castle, *S. F. Gray*. Devonshire, *Rev. J. M. Chanter*. Conway, *Dr. Allchin*. Denbighshire: Ruthin, *T. Pritchard*. The Craigs, near Dumfries, *W. G. Johnstone*. Ireland: Mucruss, *Dr. Allchin*. It is possible some of these may really belong to the less developed states of *semilacerum*, since the less perfect fronds of the latter much resemble them. [Plate III E.—Folio ed. t. III B.]

13. *multiforme* (Clowes). This is a most variable form, no two fronds being alike. In some cases the lobes are compound, approaching *semilacerum*, being deeply serrated or lobulate, with the lobules minutely serrated. In others the longitudinal development of the frond is so much arrested that it becomes truncate, or cut short, the leafy portion being sometimes partially wanting, so that the midvein, or other minor veins, form horn-like projecting points; the latter form, which has been called *truncatum*, was found in Ireland, by Dr. Allchin. The more general form it assumes is that of becoming irregularly branched, the branches proceeding indifferently from near the base or apex or centre of the frond, and the lobes being here perfect, there dwarfed or altogether wanting, or sometimes enlarged and deeply-lobed, or toothed, or having exaggerated auricles. The excurrent rib or vein is very frequently produced; and where the frond is branched, the lobes of the branches are equally affected. Sometimes a pinnatifid lobe right and left from the top of the stipes forms the entire frond. This has been sent to us from—Windermere, *F. Clowes*. Yorkshire: Coxwold, *C. Monkman*. Stirling, *Mrs. Macleod*.

14. *semilacerum* (Lk.). This, which is often known as the Irish Polypody, is the most compound of all the *fertile* forms of this species, and is permanent under cultivation. The fronds are very beautifully and symmetrically divided, the primary lobes being

themselves lobed, and in some instances, the lobules again divided or serrated. There are several slight variations: sometimes, especially in young or recently divided plants, the fronds are merely crenately-serrated, and they occur in various degrees of development. The fronds are irregularly bipinnatifid, from a foot to a foot and a-half in length, elongate-ovate in outline, pinnatifid, in the lower part almost pinnate. The primary lobes in the lower half of the frond are narrow and deeply serrate at the base and apex, deeply pinnatifid about the middle; the secondary lobes or lobules are linear, acute or bluntish, serrate, longest at the lower part of the frond, becoming shorter upwards. It is this lobed condition of the lower half of the frond which seems to have suggested the name. The veins from the principal midvein extend along the lobules, and become branched, the branches dividing into from two to three venules; while in the less divided parts, the veins are arranged similarly to those in luxuriant examples of the normal form. The upper half of the frond is fertile, and in this fertile portion the lobes are scarcely subdivided, the uppermost ones being merely serrate or crenato-serrate; the development of the lobules, and of the sori, are consequently not generally coincident on the same parts. Occasionally the whole frond, instead of the upper portion only, is merely crenato-serrate; it then agrees with the *sinuatum* of Mr. Francis. This beautiful plant, in its best state, was found many years since in a wood near the Dargle in the county of Wicklow, Ireland. Other more or less modified forms of it have been met with—Cornwall, *Rev. C. Trelawny*. Devon: Berry Pomeroy Castle, *Sir W. C. Trevelyan*; Torquay; Sidmouth. Somerset: Cheddar; Nettlecombe, *C. Elworthy*. Isle of Wight: Bonchurch. Kent: Saltwood Castle, *S. O. Gray*. Norfolk: Postwick, *Hb. Hooker*. Monmouth: Tintern Abbey, and Chepstow Castle, *R. Heward*. Pembrokeshire: Castle Malgwyn, *W. Hutchison*. Carnarvon: Aberglaslyn, *Dr. Allchin*. Lancashire: Ulverstone, *Miss E. Hodgson*. Kerry: Killarney, *J. Ball, Dr. Allchin*. Clare: Ballinahinch, *J. R. Kinahan*. Waterford: Blackwater, *J. R. Kinahan*. Antrim: Red Hall, Carrickfergus. Galway: Lough Coota, *J. R. Kinahan*. Arran Isles. Guernsey, *Dr. Allchin*. It grows, moreover, in Germany; and at Cintra in Portugal. [Plate V.—Folio ed. t. II A.]

15. *cambricum* (Lin.). This, which is perhaps the most beautiful of all known varieties of *Polypodium*, is commonly called the Welsh Polypody. It is the most compound of the varieties, quite permanent under cultivation, and under all conditions uniformly *barren*. The fronds, which are regularly bipinnatifid throughout, are ovate or ovate-oblong in outline, the lower pairs of lobes being scarcely smaller than those above them, so as to give squareness or angularity to the base. The lobes are crowded, narrow at the base, and acuminate at the apex, the intervening portion being much widened and overlapping, and the whole margin, except the very base and apex, divided into narrow linear or linear-lanceolate, acute or acuminate, serrated, crowded lobules. This variety, now chiefly known as a garden plant, was originally found in Wales, whence its name; the older books record as habitats—Glamorganshire: rocks near Dennys Powys Castle, not far from Cardiff; rocks in North Wales; and Kidderminster. It has more recently been reported to occur in Cheshire: wood near Macclesfield, *E. J. Lowe;* and in Gloucestershire: Almondsbury, near Bristol, *E. Morse*. The same variety has been found at Montpelier. [Plate VI.—Folio ed. t. III A.]

16. *omnilacerum* (M.). This is a very beautiful variety, most nearly related to *cambricum*, from which it differs in the lobes not being narrowed near the base and in bearing fructification; while from *semilacerum* it differs in being bipinnatifid throughout. The fronds are ovate oblong, a foot or more in length, bipinnatifid from the base to the apex. The primary lobes are caudate at the apex, sometimes very regularly margined with lobules, so that their basal part is nearly equal in width with the middle and upper portion, but occasionally one or two of the lower lobules are shorter, while in other instances and often in the same frond, they are longer than the rest; these lobules are not so much crowded as in *cambricum*, sometimes of unequal length, usually linear acuminate, and distinctly serrate. The fronds are sparingly fertile. It was found about Goodrich Castle, Ross, Herefordshire, by Mr. E. T. Bennett, and proves to be a constant variety under cultivation. [Plate VII.]

THE MOUNTAIN POLYPODY, or BEECH FERN.

POLYPODIUM PHEGOPTERIS.

P. fronds ovate triangular, acuminate, membranaceous, pinnate below; pinnæ lanceolate, the lower pair distinct, sessile, usually deflexed, pinnatifid; lobules linear-oblong, blunt; upper pinnæ confluent; veins hairy beneath. [Plate VIII.]

Polypodium Phegopteris, *Linnæus, Sp. Plant.* 1550. *Bolton, Fil. Brit.* 36, t. 20. *Hudson, Fl. Ang.* 456. *Smith, Fl. Brit.* 1116; *Id., Eng. Bot.* xxxi. t. 2224 and xv. t. 1018, as *Aspidium Thelypteris; Id., Eng. Fl.* 2 ed. iv. 269. *Deakin, Florigr. Brit.* iv. 41, fig. 1580. *Hooker & Arnott, Brit. Fl.* 7 ed. 581. *Babington, Man. Brit. Bot.* 4 ed. 419. *Bentham, Handb. Brit. Fl.* 626. *Lowe, Nat. Hist. Ferns,* i. t. 29. *Sowerby, Ferns of Gt. Brit.* 11, t. 2. *Moore, Handb. Brit. Ferns,* 3 ed. 56; *Id., Ferns of Gt. Brit. and Ireland, Nature Printed,* t. 4. *Mackay, Fl. Hib.* 337. *Schkuhr, Krypt. Gew.* 17, t. 20. *Willdenow, Sp. Plant.* v. 199. *Sprengel, Syst. Veg.* iv. 57. *Presl, Tent. Pterid.* 180. *Fries, Sum. Veg.* 82. *Flora Danica,* t. 1241. *Ledebour, Fl. Ross.* iv. 508. *Koch, Synops.* 2 ed. 974. *Gray, Bot. North. U. States,* 590. *Nyman, Syll. Fl. Europ.* 431.

Polypodium connectile, *Michaux, Fl. Bor. Amer.* ii. 271. *Willdenow, Sp. Plant.* v. 200.

Polypodium latebrosum, *Salisbury, Prod.* 403.

Polystichum Phegopteris, *Roth, Fl. Germ.* iii. 72.

Lastrea Phegopteris, *Bory, Dict. Class. d'Hist. Nat.* ix. 233. *Newman, Nat. Almanac,* 1844, 17; *Id., Brit. Ferns,* 2 ed. 13.

Phegopteris polypodioides, *Fée, Gen. Fil.* 243.

Phegopteris vulgaris, *Mettenius, Fil. Hort. Bot. Lips.* 83. *J. Smith, Cat. Cult. Ferns,* 17.

Gymnocarpium Phegopteris, *Newman, Phytol.* iv. 371; *Id.,* 1851, *Appendix,* xxiii.; *Id., Hist. Brit. Ferns,* 3 ed. 49.

Caudex creeping extensively, branched, tough, slender, about the thickness of a straw, dark-brown, pilose and slightly scaly while young, the older portions denuded both of scales and hairs. *Scales* lanceolate, golden-brown, intermixed with other cobwebby hair-like ones. *Fibres* numerous, much branched, dark-brown, invested with cobwebby deciduous pubescence.

Vernation circinate; the pinnæ rolled up separately towards the rachis, which is then rolled from the point downwards.

Stipes as long as, or more frequently longer, often much longer than the frond, erect, brittle, pale-green, furnished near the base with a few lanceolate acuminate pale-brown deciduous scales, and on the upper part with a few scattered subulate ones, clothed along the whole length with minute reversed hairs; distant and lateral on the caudex.

Fronds from four to eighteen or twenty inches in length, including the stipes, adherent to the rhizome, membranaceo-herbaceous, of a dull pale-green, hairy, ovate-triangular, much acuminate, pinnate below, the apical portion pinnatifid. *Pinnæ* deeply pinnatifid, linear-acuminate, nearly or quite opposite; the lower pair lanceolate, deflexed, distant from the upper, sessile, but attached only by their rachis; upper pinnæ sessile and broadly attached, and, except occasionally the second pair, confluent, so that the two basal lobules of the bases of the opposite pairs, unite to form a cruciform figure. The upper pinnæ have their points directed towards the apex of the frond, and their decurrent bases are continuous along the rachis. *Lobules* oblong-obtuse, entire, or slightly crenato-dentate, directed towards the apex of the pinnæ.

Venation of the lobules consisting of a slender flexuous *midvein*, from which proceed alternate or sometimes opposite *veins;* these veins extend to the margin of the lobule, and are either simple, or become once forked about half-way their length; the veins when simple, or when divided, the anterior *venules*, bear a sorus at a short distance from the edge of the lobule.

Fructification on the back of the frond, scattered almost equally over the whole surface. *Sori* circular, small, quite destitute of covering, arranged in a series near the margin of the lobules, and often becoming confluent in lines. Where the fructification is but partially developed, only one or two of the lowermost veins are fertile, in which case the marginal series of sori is not very manifest. *Spore-cases* small, numerous, pale-brown. *Spores* ovate, smooth.

Duration. The rhizome is perennial. The fronds are annual; produced about May, and destroyed by the early frosts of autumn.

This Fern is readily known from its congeners by its outline, which is ovate-triangular with an elongated narrow point; by the pinnato-pinnatifid mode in which its fronds are divided; by the hairiness of its surface; and by the direction of its pinnæ.

The fronds in this species become lateral and distant on the underground caudex, in consequence of its rapid elongation; and they are adherent, that is to say, their stipes is not furnished with any natural point of spontaneous separation. The character of an underground caudex seems principally relied on by Mr. Newman in establishing his genus *Gymnocarpium*, which consists of the present species, together with *P. Dryopteris* and *P. Robertianum*. Mere peculiarities of habit, however, are insufficient to mark out generic groups, and must not be permitted to override the characters afforded by the organs of fructification. In the case of the *Phegopteris* group to which our present species belongs, there is, in the characters of the fructification, so close a similarity, amounting almost to identity, with those of the typical group of *Polypodium*, that their separation is, we think, unnecessary and unwarranted. In fact, the only differential character of any importance, afforded by the fructification, is that of the medial, not terminal, position of the sori on the veins. M. Fée had already founded his genus *Phegopteris*, agreeing with Presl's section of the same name, mainly on this character, but as this feature is not everywhere constant, the genus cannot be held to be sufficiently established.

This species is rather widely dispersed, and not unfrequent throughout Great Britain, occurring most abundantly in the north and west of England and in Scotland; inhabiting shady humid places, and usually districts which are more or less hilly. In the south-eastern parts of England it appears to be wanting, but it is found rather plentifully both in Sussex and in the western counties. It occurs plentifully in Wales, and is also met with in the Hebrides, and Orkneys, in Shetland, and in the Isle of Man. In Ireland it is rare, occurring principally in the northern and eastern provinces. In elevation it extends from the coast level in the west of England, to upwards of 3000 feet in the western Highlands. The records of its distribution are as follows, the habitats being arranged in botanical districts or provinces, agreeing generally with those adopted in Mr. Watson's *Cybele Britannica*.

Peninsula.—Cornwall: near Tintagel. Devonshire: Exmoor, near Challacombe, *R. J. Gray;* Cock's Tor, *Rev. W. S. Hore;*

White Tor, Great Mist Tor, and Sheep's Tor, *R. J. Gray;* Dartmoor, *R. J. Gray;* Becky Falls, etc.

Channel.—Sussex: Kidbrood Park, Forest Row; Tilgate Forest, *S. O. Gray;* Balcombe, *J. Lloyd.*

Severn.—Gloucestershire: Ankerbury Hill, Forest of Dean; near Lydbrook. Herefordshire: Shobden Hill woods; Aymestrey quarry. Monmouthshire: Pont-y-pool, *T. H. Thomas.* Staffordshire: Ridge Hill; Madeley, etc. Shropshire: Titterstone Clee Hill; Craigforda, near Oswestry, *Rev. T. Salwey;* near Ludlow.

Trent.—Derbyshire: Buxton.

Mersey.—Cheshire: Mow Cop; Early Banks Wood near Staleybridge; Werneth, etc. Lancashire: Dean-Church Clough, near Bolton; near Todmorden; Philips Wood, near Prestwich; Boghart Hole Clough; Fox Clough, near Colne; Blackhay, near Clitheroe; Chaigeley Manor, *E. J. Lowe;* Longridge Fell; Mere Clough; and woods near Manchester, etc.

Humber.—Yorkshire: Halifax; Beckdale Helmsley; Buttercrambe Moor near York; Settle; Sheffield; Ingleborough; Bradford, *J. T. Newboult;* Kettlewell; Hebden Bridge; Teesdale; Bolton Abbey Woods; Wensleydale; and many other parts.

Tyne.—Durham: by the Tees above Middleton; rocks above Langley Ford; Cawsey Dene; Waskerley Dene, etc. Northumberland: moors near Wallington; Shewing Shields; Cheviot Hills; Hexham; Banks of the Irthing, *Rev. R. Taylor.*

Lakes.—Cumberland: Wasdale; Borrowdale; Ennerdale; Scaw Fell; Keswick; Tindal Fell; Newbiggin Wood; Laggat, on Cold Fell, *J. Robson,* etc. Westmoreland: Stockgill Force; Ambleside; Grassmere; Casterton Fell; Cheviot Hills; Wallington; Hutton Roof, etc. N. Lancashire: Conistone. Isle of Man.

S. Wales.—Radnorshire: Craig-Pwll-du; Rhayader. Brecknockshire: Ystrad-y-Ffin, *T. H. Thomas;* Uscoed Hendry; Uscoed Eynon Gam; Pont Henryd near Capel Colboen; Brecon Beacon, etc. Glamorganshire: Pont Nedd Vechn; Scwd-y-Gladis; Cilhepste. Carmarthenshire: Glynhir, near Llandebio. Cardiganshire: Devil's bridge; Hafod, etc.

N. Wales.—Montgomeryshire: Garthbeibio; Plinlymmon. Merionethshire: Falls of the Cynvael near Festiniog; Rhaidr-y-Mawddach;

Barmouth, etc. Denbighshire: Llanrwst; Ruthin; Rhuabon, *A. L. Taylor*. Carnarvonshire: Cwm-Idwal; Dolbadern; Llanberis; Rhaidr-y-Wenol; Beddgelert; Twll-du; Capel Curig; Aberglaslyn; Bangor, etc.

W. Lowlands.—Dumfriesshire: Drumlanrig; Rae Hills; Jardine Hall; Dumfries. Kircudbrightshire: Dalscairth; Mabie, *P. Gray*. Renfrewshire: Gourock. Lanarkshire: Falls of the Clyde, near Corra Linn; Calderwood; Crutherland; Campsie near Glasgow, etc.

E. Lowlands.—Berwickshire. Roxburghshire: Jedburgh; Rubers-law. Edinburghshire: Pentland Hills; Arniston; Rosslyn, and Auchindenny Woods.

E. Highlands.—Stirlingshire: Ben Lomond, *J. S. Henslow*. Clackmannanshire: Castle Campbell, near Dollar, *J. T. Syme*. Fifeshire: Dunfermline; Inverkeithing; Carden Den. Kincardineshire. Perthshire: Glen Queich in the Ochils; Bridge of Bracklin, near Callender; Dunkeld, *A. Tait;* Ben Voirlich; Ben Lawers; Craig Chailliach; Corrach Uachdar, Glen Lochy; Killin; Tyndrum; Dalnacardoch, etc. Forfarshire: Canlochen, Clova. Aberdeenshire: Castleton, Braemar. E. Inverness-shire: Dalwhinnie. Morayshire.

W. Highlands.—W. Inverness-shire: Aberarder; Ben Nevis; Red Caird Hill, etc. Argyleshire: Glen Gilp, Ardrishiag; Dunoon; Oban; Crinnan; Inverary; Balachallish; Pass of Glencroe, etc. Dumbartonshire: Tarbet; Arroquhar, etc. Isles of Mull, Islay, and Cantyre.

N. Highlands.—Ross-shire: Kessock. Sutherlandshire: Ferry-house E. of Loch Erbol. Caithness: Morven, rare, *T. Anderson.*

N. Isles.—Orkney: Hoy, *T. Anderson*. Shetland: North Marm.

Ulster.—Antrim: by the Glenarve, near Cushendall, and other parts. Donegal: waterfall above Lough Eske. Down: Slieve Bignian; near Slieve Croob; Black Mountain, above Tollymore Park. Londonderry: Glen Ness.

Connaught.—Galway: Garoom Mountain, Letterfrach, Connemara, *E. T. Bennett.*

Leinster.—Louth: Carlingford Mountain. Wicklow: Powerscourt waterfall.

Munster.—Kerry: between Killarney and Kenmare; Mucruss.

This Fern is scattered nearly throughout Europe, extending from Iceland and the Scandinavian countries southwards through the British Isles and continental Europe—France, Belgium, Holland, Switzerland, Germany, Hungary, Croatia, and Transylvania, to Spain, Northern Italy, and Greece. In Asia it is recorded from Unalaschka and Kamtchatka, and also as extending along the chain of the Altai. Algeria, in Africa, is said to produce it; whilst in America, where it is sometimes known under the name of *P. connectile,* it is met with from Greenland and Labrador on the eastern side, to Prince William's Sound on the western, extending southwards to the Rocky Mountains, to Canada, and to the northern United States.

In cultivation the Mountain Polypody requires a free supply of water; and at the same time, in order that this supply may not stagnate about its roots, very perfect drainage should be provided. This is best done by using broad shallow pots, and filling up about two-thirds of their depth with coarse rubbly materials, to allow of the percolation of the water, which, moreover, should not be too continuously kept in feeders about the bottoms of the pots. Turfy peat mixed with leaf-mould in the proportions of two-thirds of the former to one-third of the latter, and the whole well blended with sand, forms a good compost. The plants are hardy enough to endure cold, but the beauty of the fronds, except in very favourable situations, can only be secured by keeping them, at least during the growing season, in some place of shelter, of which none can be more congenial to the plants than a cold shady frame, or its equivalent. The same remarks apply to *Polypodium Dryopteris.*

The Mountain Polypody is not liable to much variation. The only abnormal form which has been observed has some of the pinnæ or pinnules bifid or multifid, and occasionally the apex of the frond is similarly divided. It also generally happens that where dichotomous division takes place, the approximate portions are at the same time depauperated. Depauperation, caused by the punctures of an insect at an early stage of development, sometimes occurs, and might be mistaken for a natural variation.

THE ALPINE POLYPODY.

POLYPODIUM ALPESTRE.

P. fronds lanceolate, herbaceous, glabrous, sub-erect, bipinnate; pinnæ narrow lanceolate from a broad base, spreading or ascending; pinnules ovate-oblong, or subfalcately ovate-lanceolate, pinnatifid; segments oblong, bluntish, serrate; stipes short; secondary rachis narrowly winged; (sori rarely spuriously indusiate). [Plate IX.]

POLYPODIUM ALPESTRE, *Hoppe, Pl. Exs. Sprengel, Syst. Veg.* iv. part 2, 320. *Kaulfuss, Flora,* 1829, 328. *Koch, Synops.* ed. 2, 974. *Moore, Handb. Brit. Ferns,* 3 ed. 59; *Id., Ferns of Gt. Brit. and Ireland, Nature Printed,* t. 7 A—C. *Henfrey, Franc. Anal. Brit. Ferns,* 5 ed. 28, *supp. plate,* fig. 2 A. *Hooker & Arnott, Brit. Fl.* 7 ed. 582. *Sowerby, Ferns of Gt. Brit.* 84, t. 49. *Bentham, Handb. Brit. Fl.* 626. *Lowe, Nat. Hist. Ferns,* i. t. 39.

POLYPODIUM RHÆTICUM, *Pallas, "Itin.* ii. 28." *Ledebour, Fl. Ross.* iv. 510. *Fries, Sum. Veg.* 82. *Woods, Tour. Fl.* 423. *Nyman, Syll. Fl. Europ.* 430; not of *Linnæus.*

ASPIDIUM ALPESTRE, *Hoppe, Taschenb.* 1805, 216. *Swartz, Syn. Fil.* 421. *Schkuhr, Krypt. Gew.* 58, t. 60 (excl. syn. Lin.).

ASPIDIUM RHÆTICUM, *Swartz, Schrad. Journ. Bot.* 1800, ii. 41; *Id., Syn. Fil.* 59 (excl. syn. Lin.). *Tenore, Att. Accad. del R. Inst. Sc. Nat. Napol.* v. (reprint 29, t. 4, fig. 8).

ASPIDIUM DISTENTIFOLIUM, *Tausch;* according to Steudel.

ATHYRIUM ALPESTRE, *Nylander;* according to Ledebour.

PSEUDATHYRIUM ALPESTRE, *Newman, Phytol.* iv. 370, 974; *Id., Appendix,* 1851, xiv.; *Id., Hist. Brit. Ferns,* 3 ed. 199. *Babington, Man. Brit. Bot.* 4 ed. 424.

PHEGOPTERIS ALPESTRIS, *Mettenius, Fil. Hort. Bot. Lips.* 83. *J. Smith, Cat. Cult. Ferns,* 16.

Var. flexile; fronds slender, flaccid, narrow lanceolate, bipinnate; pinnæ short, ovate-lanceolate, spreading or deflexed; pinnules oblong, obtuse or acutish, narrowed below, sessile or adnate, distantly lobed or toothed; stipes very short. [Plate X.]

POLYPODIUM ALPESTRE, *var.* FLEXILE, *Moore, Ferns of Gt. Brit. and Ireland, Nature Printed,* t. 7 D—E; *Id., Handb. Brit Ferns,* 3 ed. 59, 61.

POLYPODIUM FLEXILE, *Moore, Handb. Brit. Ferns,* 2 ed. 225. *Henfrey, Franc. Anal. Brit Ferns,* 5 ed. 29, *supp. plate,* fig. 2 B.

PSEUDATHYRIUM FLEXILE, *Newman, Phytol.* iv. 974; *Id., Hist. Brit. Ferns,* 3 ed. 203. *Babington, Man. Brit. Bot.* 4 ed. 424.

ATHYRIUM? FLEXILE, *Moore MS., in Hb.*

Caudex short, erect or decumbent, consisting of the persistent crowded bases of the fronds attached around a central axis, the whole forming a stout roundish mass, frequently tufted, scaly above. *Scales* numerous, broadly or narrowly ovato-lanceolate, of a pale-brown colour. *Fibres* stout, branched, dark-coloured.

Vernation circinate.

Stipes short, from about one-sixth to one-fourth of the entire length of the frond, stoutish, swollen near the base, clothed sparingly with ovate-lanceolate pale-brown scales; terminal and adherent to the caudex. *Rachis* stout, rounded behind, channelled in front; the rachis of the pinnæ furnished with a very narrow leafy wing on both sides, connecting the pinnules.

Frond from one to three feet and upwards in height, erect or ascending, herbaceous, dark dull green, lanceolate or oblong-lanceolate, the base narrowed in about the same degree as the point; bipinnate or subtripinnate. In fronds, of which the leafy portion measures about twenty inches in length, the greatest breadth is about six and a half inches. *Pinnæ* broadly linear or lanceolate from a broad base, tapering to a narrow point, numerous, crowded above, more distant below, spreading or somewhat ascending. *Pinnules* ovate-oblong, sometimes ovato-lanceolate, or oblong-ovate, acute, with a narrow attachment at the base, but connected by a narrow membranous wing which borders the rachis; they are deeply pinnatifid, and in the most vigorous fronds so much so, and the segments so far distant from each other, as to appear again pinnate. *Segments* oblong obtuse, sharply serrate, especially at the apex and on the anterior margin. The subtripinnate fronds have the segments doubly toothed.

Venation of the pinnules consisting of a slightly flexuose midvein from which branch a series of alternate pinnate veins. *Veins* of the segments flexuose, with simple alternate *venules*, one of which is directed to the point of each marginal tooth; the lowest anterior venule, which is directed towards the lowest anterior tooth, is usually soriferous, and when this only is so, the sori form a series on each side the midvein, at a short distance from it, and just above the sinus of the segments on their anterior margin; sometimes, however, some of the other venules are also fertile, and the sori are then placed near the margin of the segments. In the subtripinnate fronds, which have the segments more or less doubly toothed, the venules are occasionally forked, the anterior veinlet, or sometimes both, bearing a sorus; in these examples the sori, three or four on each side the segment, form tolerably distinct submarginal lines. The sori are in all cases attached near to, but below, the apex of the vein, which reaches to the margin.

Fructification on the back of the frond, occupying the upper two-thirds of its length. *Sori* small, circular, usually distinct, but sometimes crowded, and becoming confluent; usually naked, but sometimes (though rarely and only in abnormal-looking sori) the spore-cases are somewhat lateral, and a membrane, which appears to be an abnormal development of the dilated portion of the vein which forms the receptacle, is produced, simulating an abortive or spurious indusium. *Spore-cases* roundish-obovate, brown, numerous. *Spores* roundish or oblong, somewhat muriculate.

Duration. The caudex is perennial. The fronds are annual, growing up in April or May, and perishing early in autumn.

The Fern is at once distinguished among the British Polypodies by its short thick erect tufted caudex, and by the lanceolate form, and bipinnate or tripinnate mode of division of its fronds. It has certainly a general resemblance to *Athyrium Filix-fœmina*, with which it appears to have been very generally confounded, but the fructification, as usually borne, is very different, and even the resemblance of the frond to that species is not found on comparison to be so close as a first glance suggests.

The short massive caudex with terminal adherent fronds, would lead those botanists who derive generic distinctions from the mode in which the plant is developed, to separate the present species from *Polypodium*. Of those who do so, some refer it to the genus *Phegopteris*, in company with the other British species we retain in the *Phegopteris* section of *Polypodium*. Mr. Newman, however, creates for it a new genus, which he calls *Pseudathyrium*. We think it may be safely retained in *Polypodium*, as here understood.

The supposed 'indusia,' ascribed to this plant, which may be noticed both in the species itself, and in the variety *flexile*, are only occasional, or even rare, and they appear never to occur in company with the more perfect sori, but only where the spore-cases are much fewer in number than usual. To us they have the appearance of lacerated membranacco-filamentous expansions of those points of the veins which form the receptacles; and appear to arise from some abnormal condition, perhaps inherent, which limits the power of producing spore-cases to the side or base of the receptacle, while on the upper side its cells are directly prolonged into the indusioid membrane. In no case have we seen what could be considered as a perfect and fully developed indusium. On the other hand, Mr. Rylands, of Warrington, who regards the plant as an *Athyrium*, has communicated the result of some observations made in 1855, in company with Mr. Wilson, as follows:—"In those sori which were large and fully ripe, the indusium could not be seen, though I imagine dissection would have shown traces of it. One sorus was found still closed, the spore-cases little developed; it was reniform, and lay alongside the venule. In many of the smaller sori remains of an indusium was seen, and in two or three it was as nearly perfect as one may expect to find it. The margin was laciniated with fine projecting points. The laciniated margins are produced by the rupture of the cuticle, and the fine points are the cell-walls thereof. The indusium is very tender, shrivels, and where the spore-cases are numerous, is speedily concealed or perhaps displaced by them; it is smaller than in the other forms of *Athyrium*. These peculiarities seem to result from the rupture of the cuticle taking place early in the progress of development of the sori; but that it has the true indusium of an *Athyrium* I think cannot be further disputed." Subsequently, in reply to a suggestion that the supposed indusium in these plants was not like the indusium of a true *Athyrium*, and was not developed at all where the sori seemed most perfect, Mr. Rylands wrote:—"The 'indusia' of *alpestre* are not, I think, confined to the imperfect sori, though after bursting they soon shrivel and disappear in the larger ones. I have compared it with *A. Filix-fœmina molle*, and though in texture, position, and general character, there was little difference, I am compelled to admit that while in the case of *alpestre* the spore-cases seemed to lie within the proper cuticle of the frond, the evidence of a distinct membrane was much clearer in *molle*. This supports your view to some extent; but, all things considered, is it sufficient to remove the plant from others so evidently its allies?" The majority of the sori, indeed all of them, with few exceptions, and those exceptions generally, if not always, having strongly marked imperfect or abnormal characters, appear to us to be the round naked masses of *Polypodium*, so that we have no alternative, repudiating as we do its separation on characters derived merely from habit and resemblance, but to retain this plant in that genus, a course in which many at least of our best botanists coincide.

This plant has been as yet, so far as relates to the United Kingdom, found only in the Highlands of Scotland, where, in the mountainous districts of the counties of Forfar, Aberdeen, and Perth, as we learn from memoranda communicated by Dr. G. Lawson, it is one of the most abundant of Ferns, exceeded perhaps only in frequency by *Lastrea montana;* in the Clova district, descending to about 2000 feet above the sea, perhaps lower, and associated with *Athyrium;* and in Aberdeenshire, ascending nearly or quite to 4000 feet. Mr. James Backhouse, jun., also states that in the Clova mountains, it occurs in company with *Athyrium*, at from 2000 to 3000 feet elevation, above which, at from 3000 to 4000 feet, the latter disappears, and *P. alpestre* becomes abundant. Mr. H. C. Watson, by whom it was found so long since as 1841, but undetermined, describes the localities in which he gathered it thus*:— "In July 1841 I gathered two fronds of this Fern in the great corrie of Ben Aulder, a lofty mountain situate on the west side of Loch Erricht, Inverness-shire. Another frond of the same species was picked at some other spot in the neighbourhood of Loch Erricht, probably on the hills between Ben Aulder and the north end of the lake, but it might be on the hills of Drumochter Forest, eastward of the lake, and if the latter, the station would be within Moray, or eastern Inverness. In 1844 I brought a frond from Canlochen Glen in Forfarshire. These specimens (except the second from Ben Aulder given to Mr. Babington) remained in my herbarium until 1851, first doubtfully labelled, and then temporarily forgotten. Their close resemblance to small fronds of *Athyrium*

* Watson, in *Cybele Britannica*, iii. 253.

Filix-fœmina made me feel very uncertain whether they could be properly referred to *Polypodium*, until Mr. Newman, to whom the Canlochen frond was at length shown when again recollected, decided it to be *P. alpestre*." It seems widely dispersed throughout the Highlands, and may even occur further southward, on the high mountains of Wales and the North of England. On the hill sides in exposed places, the fronds are very commonly damaged either by winds, spring frosts, or by animals, and it is only in the more sheltered localities that perfect specimens can be obtained. The recorded habitats for this species are the following:—

E. Highlands.—Perthshire: Killin; Ben Lawers. Forfarshire: Glen Fiadh, Glen Prosen, Glen Dole, Canlochen, Glen Isla, and other glens of the Clova Mountains, *J. Backhouse, G. Lawson.* Aberdeenshire: Braemar; by the streams on Ben-Aven, Ben-na-Bourd, Ben-Mac-d'hui, and by the lake which forms the source of the Dee, *A. Croall;* Glen Callater; Cairntoul; Loch-na-gar. In all the Corries of the Dee-side mountains, and those of the neighbouring districts, often mixed with *Athyrium Filix-fœmina* at an elevation of from 2000 to 3000 feet; from 3000 to 4000 feet *Filix-fœmina* had ceased, while *P. alpestre* was plentiful; in damp gorges and among tumbled rocks, often destitute of fructification, but in exposed places abundantly fructified, *J. Backhouse.* Abundant in the mountains of Aberdeen, Forfar, and Perth, at from 2000 to 4000 feet elevation, *G. Lawson.* Mountains near Dalwhinnie, east side of Inverness-shire, 1841, *H. C. Watson.*

W. Highlands.—Great Corrie of Ben Aulder, west side of Inverness-shire, 1841, *H. C. Watson.*

N. Highlands.—Sutherlandshire: Ben Hope.

In Europe this plant is met with in alpine and subalpine situations in Norway, Sweden, and Lapland; in Russia, in Livonia on the west and the Ural mountains on the east; in Scotland, and in France and Belgium; in the Alps of Switzerland, in Germany, and in Spain. Meyer found it in the Caucasus. A scarcely distinguishable plant, perhaps identical, was collected by Barclay at Sitka, in North-West America, as appears from specimens in Sir W. J. Hooker's herbarium.

We have seldom seen cultivated plants of this species thriving with the vigour they possess in their native hills, except when grown fully exposed to the air in sheltered shady situations and in a pure atmosphere. When confined within doors, especially in smoky localities, they often produce but puny and flaccid fronds. It roots, however, freely in a sandy compost of loam and peat, and with a free admission of air may be grown with tolerable success in frames or cool fern houses where it is necessary to use these means of sheltering plants of this nature from atmospheric impurities. In all cases, however, where a pure atmosphere is enjoyed, it will be found to grow better on the open rockery, than in pots under glass; and when it is found requisite to adopt frame or house culture, the plants, though shaded, should be provided with as airy and light a situation as can be found. In situations where it can be grown out-doors the species is perfectly hardy, of erect habit, and not inelegant, though by no means comparable in beauty with some states of the Lady Fern, which it most resembles. The variety *flexile* forms a beautiful pot plant for a cool house or frame, its delicate spreading and gracefully curving fronds giving it a character of distinctness as well as elegance. The plants require good drainage, as they like a tolerable supply of water; and though they must have moderate shade they are not benefited by being placed in a confined situation. They may be increased by separating the lateral crowns of the caudex, as well as like most other Ferns, from the spores.

This Polypody is a very variable plant, but we have as yet no experience whether the variations are generally such as would be perpetuated under other conditions than those in which they naturally occur. The forms we have received from the natural habitats have a remarkable correspondence, as regards their general character and division, with those of *Athyrium Filix-fœmina*. The most striking of them are enumerated as sub-varieties below, with the object of recording, as we have done in other cases, the most marked modifications of development to which the species is subject.

1. *flexile* (M.). This is a narrow lax form, with deflexed pinnæ; and bears perhaps in its irregular toothing, and singular habits of

fructification, some indication of being a monstrous or abnormal variation, though it is perfectly constant to the peculiarities above assigned to it. It differs from *P. alpestre* in being more slender and flaccid; in having a much narrower outline, and consequently shorter pinnæ, with a considerably reduced number of pinnules; in the form of the pinnules, which are oblong, narrowed below, sessile or adnate, and distantly toothed; in the very short stipes which becomes obsolete in the cultivated plants; and in a tendency to bear perfect sori at the base of the frond, while the apex is barren—the reverse of what usually happens. The fronds are from six or seven to twelve or eighteen inches in length; the pinnæ, spreading or more or less deflexed, short, with about six or eight pairs of pinnules. The sori are few, six or eight on a pinnule, usually distinct; in the cultivated plant the clusters are very numerous in the lower half, and scarcely extend upwards beyond the middle of the frond; but this character is not constant, the fronds being sometimes fructified throughout, and sometimes fertile both at the base and apex. The spore-cases sometimes appear for the most part to be attached to the side of the vein, and the sori slightly elongated rather than circular, indicating an affinity with *Athyrium;* and there is in some cases a peculiar membranaceo-filamentous development in the position of an indusium, again indicating affinity with the ciliated indusia of *Athyrium;* but at the base and apex of the frond, the more perfect sori are generally without trace of this indusioid growth, and truly polypodioid. The absence of stipes, which Mr. Newman includes in his definition, is not constant, the wild specimens sent by Mr. Backhouse having a distinct stipes about a couple of inches long; this part, however, is always short. It is certainly a very distinct variety, and very constant, probably a variety rather than a species, this, moreover, being the view adopted by its discoverer, Mr. Backhouse, who writes:—"Dissimilar as it is from *P. alpestre,* I shall continue doubtful of its specific difference if it does not turn up in other places." Mr. Backhouse, by whose party only it has been found wild, met with it in one place only, but in some quantity, in Glen Prosen, Clova, Forfarshire.

2. *lanceum* (M.). In this the fronds are large, stout, subtripinnate; the pinnules elongate, ovato-lanceolate or sometimes sublinear,

slightly falcate, deeply pinnatifid, with obtuse serrated segments, the lowest of which are almost separate. We have received it from Dr. G. Lawson, and Mr. Croall, gathered at the White Water Falls and elsewhere in the Clova mountains; Mr. Croall has also communicated the same form from Loch-na-gar, Aberdeenshire.

3. *tripinnatum* (M.). The fronds of this form are large, stout, tripinnate; the pinnules, which are from an inch to an inch and a-half long, are oblong-ovate, with separate, oblong, secondary pinnules, the upper of which are united by the wing of the rachis, but the lower are separate to their base. It is analogous to fine states of *Athyrium Filix-fœmina incisum.* Dr. G. Lawson gathered it at the Wells of Dee, Aberdeenshire.

4. *laciniatum* (M.). An elegant variety raised from spores of the species by Messrs. Stansfield and Son, of Todmorden, in 1857. It is analogous to the laciniate varieties of the Lady Fern, having the pinnules irregularly depauperated, or jagged, or confluent, the pinnæ themselves, as regards their length, not being much affected in the specimens we have seen.

The smaller and more usual, at least the more usually collected, forms of this plant, are analogous to the smaller forms of *Athyrium Filix-fœmina;* even these, however, exhibit differences in habit, some being quite erect, while others are spreading, as in the Lady Fern.

We suspect that a dwarf barren monstrous shy-growing plant found by Dr. Dickie on Ben-Mac-d'hui, and hitherto referred to *Athyrium Filix-fœmina* (var. *præmorsum*), belongs rather to this species. It has been thus described in our folio edition:—"This curious dwarf, and as yet barren form, was found by Dr. Dickie on 'Ben-na-Muich-dhu', at an altitude of 3700 feet, in 1846, and has since that time proved constant under cultivation. The fronds which rarely attain a height of eight inches, are of an irregular ovate-lanceolate outline. The pinnæ are unequal, and the pinnules are oblong and decurrent, lacerate, and irregular as if they had been partially eaten by an insect. It is exceedingly rare."

THE SMOOTH THREE-BRANCHED POLYPODY, or OAK FERN.

POLYPODIUM DRYOPTERIS.

P. fronds pentangular-deltoid, ternate, smooth, membranaceous; branches pinnate; pinnæ deeply pinnatifid (sometimes pinnate at the base); lobules or pinnules oblong, obtuse, crenate or crenately lobate; stipes glabrous. [Plate XI.]

POLYPODIUM DRYOPTERIS, *Linnæus, Sp. Plant.* 1555. *Bolton, Fil. Brit.* 52, t. 28. *Smith, Fl. Brit.* 1116; *Id., Eng. Bot.* ix. t. 616; *Id., Eng. Fl.* 2 ed. iv. 269. *Hudson, Fl. Ang.* 460. *Deakin, Florigr. Brit.* iv. 42, fig. 1581. *Hooker & Arnott, Brit. Fl.* 7 ed. 582. *Babington, Man. Brit. Bot.* 4 ed. 420. *Sowerby, Ferns of Gt. Brit.* 11, t. 3. *Moore, Handb. Brit. Ferns,* 3 ed. 64; *Id., Ferns of Gt. Brit. and Ireland, Nature Printed,* t. 5. *Mackay, Fl. Hib.* 338. *Bentham, Handb. Brit. Fl.* 626. *Lowe, Nat. Hist. Ferns,* i. t. 27. *Schkuhr, Krypt. Gew.* 19, t. 25. *Willdenow, Sp. Plant.* v. 209. *Sprengel, Syst. Veg.* iv. 60. *Presl, Tent. Pter.* 180. *Fries, Sum. Veg.* 82. *Koch, Synops.* 2 ed. 974. *Gray, Bot. North. U. States,* 590. *Flora Danica,* tt. 759, 1943. *Sturm, Deutschl. Fl. (Farrn.)* t. 7.

POLYPODIUM DRYOPTERIS, α. GENUINUM, *Ledebour, Fl. Ross.* iv. 509.

POLYPODIUM PULCHELLUM, *Salisbury, Prod.* 403.

POLYSTICHUM DRYOPTERIS, *Roth, Fl. Germ.* iii. 80.

LASTREA DRYOPTERIS, *Bory, Dict. Class. d'Hist. Nat.* ix. 233. *Newman, Nat. Alm.* 1844, 15; *Id., Brit. Ferns,* 2 ed. 13.

PHEGOPTERIS DRYOPTERIS, *Fée, Gen. Fil.* 243. *J. Smith, Cat. Cult. Ferns,* 17.

GYMNOCARPIUM DRYOPTERIS, *Newman, Phytol.* iv. 371; *Id.,* 1851, *Appendix,* xxiv.; *Id., Hist. Brit. Ferns,* 3 ed. 57.

Caudex creeping extensively, branched, tough, slender, about the thickness of a straw, dark-brown, almost black, the younger portions scaly. *Scales* lanceolate, like those of the stipes, pale semi-transparent brown. *Fibres* dark-brown, branched, clothed with fine pubescence.

Vernation circinate; the lateral or lower pair of branches rolled up separately from the remaining central portion, so that the young fronds resemble, as Mr. Newman expresses it, three little balls set on slender wires at the top of the stipes.

Stipes very much longer than the fronds, frequently twice or thrice their length, erect, slender, brittle, tinged with purple, and furnished near the base with a few scattered pale-brown lanceolate deciduous scales, otherwise smooth and glabrous; lateral and adherent to the caudex, and somewhat distant. *Rachis* quite smooth; that of the central branch deflexed, of the lateral branches spreading.

Fronds from four to twelve or fourteen inches in height, including the stipes, the leafy portion averaging from four to six inches in length and breadth, delicately membranaceous, bright lively green, quite smooth; in form deltoidly-pentangular, the pentagon being described by the points of the three branches and those of the two basal pinnules of the lower branches, which latter diverge so as to represent two separate angles. The fronds are ternate, that is, they consist of three nearly equal portions or branches, as indicated by the vernation. *Branches* pinnate or subbipinnate, differing from each other chiefly in this, that while the upper or central one has its sides nearly equal, the two lateral ones have the pinnæ on their lower side larger, sometimes twice as large as those on the upper side, so that they are obliquely triangular. *Pinnæ* opposite, variable in outline from ovate to linear-oblong, acute, usually pinnate at the base, pinnatifid above, and acute as well as nearly entire at the apex; those of the central branch more decidedly pinnate than those of the lateral ones. *Pinnules* or *lobules* oblong-obtuse, crenate or crenato-lobate, smaller and less divided towards the apex.

Venation of the more compound, that is the crenato-lobate pinnules, consisting of a flexuose *costa* or midvein with alternate *veins*, one to each lobe, these veins being pinnato-furcately branched, with the *venules* extending to the margin. The veins of the crenate pinnules have fewer branches or venules. The first anterior venule bears a sorus some distance below its termination.

Fructification on the back of the frond, and spread over its whole surface. *Sori* small, circular, consisting of numerous crowded spore-cases, quite uncovered, arranged in a linear, often crowded, series, along each side of the pinnules, near to but distinctly within the margin, the sori being seated some distance below the apex of the venules. Sometimes the fronds are less abundantly fructified, and the sori appear distant and scattered. *Spore-cases* small, dark-brown, roundish-obovate, attached by a slender pedicel. *Spores* ovate, roundish, or oblong, with a granulated surface.

Duration. The caudex is perennial. The fronds are annual,

produced about April, and in succession through the summer, and perishing early in autumn.

The nearest affinity of this species is with *P. Robertianum*, from which some botanists do not think it distinct. It can, however, hardly be supposed that those who have seen tolerably good living examples of both kinds, can adopt this opinion. *P. Dryopteris* differs from *P. Robertianum* in having a loosely spreading habit, while the fronds of the latter are rigid and erect, with stouter stalks and ribs, and a less membranaceous texture; it differs further in having ternate or three-branched fronds, which is not strictly the case with the latter, although by a misapplication of terms it is sometimes so described. *P. Dryopteris* is decidedly three-branched, as its vernation, compared to three little balls on slender wires, certifies; whilst in *P. Robertianum*, as Mr. Newman well states, the three corresponding portions of the frond never assume this appearance, but on the contrary, every pinnule is rolled up into a little globe, the pinnæ rolled in on their rachides, and the entire frond upon its rachis, so that the frond is of the ordinary bipinnate structure. Of less botanical importance perhaps, but equally, or still more clearly available as a distinguishing characteristic, is the perfect smoothness of *P. Dryopteris*, compared with the glandular pubescence of *P. Robertianum*, most readily seen on the stipes and rachis, but equally occurring over the whole plant. These peculiarities, which are perfectly constant in a state of cultivation, mark the plants as abundantly distinct.

Most writers describe a cruciform figure as being formed by the basal pinnules of the opposite sessile pinnæ in *P. Dryopteris;* and it is sometimes figured, as in Mr. Newman's work, in a very marked manner. Some approach to this arrangement is indeed at times observable, but even extensive suites of specimens may fail to show it in any remarkable degree; and when it does occur, two of the four pinnules (the upper pair) are smaller, and nearly parallel, while the lower and larger ones are divergent.

This Fern, though not a common plant, is widely dispersed in Great Britain. It occurs principally in wild and mountainous rocky districts, in the neighbourhood of waterfalls, and in the drier parts of wet woods; sometimes growing in limestone districts along with *P. Robertianum*. Its north limit is attained in Sutherlandshire; hence it extends southwards, through the Lowlands of Scotland and the north of England, to South Wales and the centre of England, occurring in Devonshire near the extreme south-west, but avoiding the eastern side of the island. It is very rare in Ireland. As regards elevation it extends from near the sea level in Devonshire, to a height of about 2700 feet in the West Highlands. Its habitats are as follows:—

Peninsula.—Devonshire: near Ilfracombe, *Rev. J. M. Chanter;* Challacombe, Exmoor, *H. F. Dempster.* Somersetshire: Mendip Hills; near Bristol; near Bath.

Channel.—? Hampshire: Petersfield, *Dr. Bromfield.* Sussex: Tilgate Forest, *Rev. T. Rooper.*

Thames.—Oxfordshire: Cornbury Quarry. ? Essex: Chingford Church.

Severn.—Warwickshire: Berkswell. Gloucestershire: Lea Bailey; New Weir; Frocester Hill; Atterbury Hill, above Lydbrook, Forest of Dean, *E. T. Bennett.* Monmouthshire: Tintern Abbey; Pont-y-pool, *T. H. Thomas.* Herefordshire: Penyard Park Wood, near Ross; Shobden Hill Woods; near Downton Castle, by the Teme; Aymestrey Quarry. Worcestershire: Malvern Hills; Shrawley Wood. Staffordshire: Trentham Park; near Colton Hall, and Oakamoor; near Yoxhall Lodge, Needwood Forest. Shropshire: Titterstone Clee Hill; Whitcliffe, near Ludlow; Froddesley Hill.

Trent.—Derbyshire: Chinley Hill, near Chapel-en-le-Frith; Pleasley Forges. Lincolnshire.

Mersey.—Cheshire: Hill Cliff. Lancashire: Manchester; Warrington; Egerton Moor, and Dean-Church Clough, near Bolton; Prestwich Clough; Boghart Hole Clough; Broadbank, near Colne; Mere Clough; Cotteril Clough; Chaigeley Manor, *E. J. Lowe;* Lancaster; Ashworth Wood; Longridge Fell, *E. J. Lowe,* etc.

Humber.—Yorkshire: Banks of the Wharfe, Burley; Brimham Rocks; Thirsk; Ingleborough; Rivaulx Woods; Teesdale; Halifax; Whitby; Richmond; Bradford, *J. T. Newboult;* Tyersal, near Bradford, *G. W. Jackson;* Sheffield; Hebden Bridge; Bolton

Abbey Woods; Settle, *J. Tatham;* Brierley; Castle Howard Park, and many other parts.

Tyne.—Durham: Walbottle Dene; foot of the Cheviots, near Langley Ford; higher part of the Tees. Northumberland: Banks of the Wansbeck; Morpeth; Hexham; Shewing Shields; Scotswood Dene; Banks of the Blythe and Irthing, *Rev. R. Taylor.*

Lakes.—Cumberland: Lodore, near Keswick; Honister Crags; Borrowdale; Calder Bridge; Wasdale; Scale Force; Dalegarth; Gillsland. Westmoreland: Stockgill Force, Ambleside; Hutton Roof; Casterton, etc. N. Lancashire: Conistone.

S. Wales.—Radnorshire: Craig-Pwll-du. Brecknockshire: Brecon; Trecastle; Pont Henryd, near Capel Colboen; Ystrad Felltree; Ystrad-y-Ffin, and Llanwyrtyd, *T. H. Thomas.* Glamorganshire: Pont Nedd-Vechn; Scwd-y-Gladis; Cilhepste; Merthyr-Tydvil. Cardiganshire: Ponterwyd; Devil's Bridge; Hafod, *J. Riley,* etc.

N. Wales.—Anglesea. Denbighshire: Llangollen; Ruthin. Montgomeryshire: Craig-Breidden; Plinlymmon. Merionethshire: Dolgelly, *A. Irvine.* Flintshire: near St. Asaph. Carnarvonshire: Cwm-Idwal; Llanberis; Bangor; Rhaiadr-y-Wenol; Twll-du.

W. Lowlands.—Dumfriesshire: Drumlanrig; Rae Hills; Maiden Bower Craigs; Langholm. Kirkcudbrightshire: Cluden Craigs; Hills above Dalscairth, *P. Gray.* Lanarkshire: Falls of the Clyde, *H. M. Balfour;* Banks of the Kelvin; Calderwood, *T. B. Bell.* Renfrewshire: Gourock.

E. Lowlands.—Roxburghshire: Wanchope, *W. Scott.* Berwickshire: Banks of the Whiteadder; Longformacus. Edinburghshire: Hawthornden; Rosslyn; Auchindenny Woods, and elsewhere about Edinburgh.

E. Highlands.—Clackmannanshire. Kinross-shire. Fifeshire: Carden Den, *R. Maughan.* Perthshire: Culross; Ben Lawers; Corrach Uachdar, Glen Lochy; Killin; Dalnacardoch; Killicrankie, *H. B. M. Harris;* Dunsinane Wood; Dunkeld, *S. O. Gray;* Pass of Trosachs; Ben Voirlich. Forfarshire: Sidlaw Hills; Clova Mountains; Glen Isla; Craig; Clack of the Balloch, *L. Carnegie.* Kincardineshire: Inglis Maldie, *A. Croall.* Aberdeenshire: Castleton, Braemar. Nairnshire: Cawdor Woods, *J. M'Nab.* E. Inverness-shire: Dalwhinnie; Glen Marson, Aberarder. Moray.

W. Highlands.—W. Inverness-shire: Freuch Corrie, Strath Affaric; Glen Roy; Ben Aulder. Dumbartonshire: by Loch Lomond. Argyleshire: Glen Gilp, Ardrishiag; between Loch Awe and Loch Etive; Dunoon. Isle of Arran: Brodick. Isle of Mull: Tobermorey, *W. Christy.*

N. Highlands.—Ross-shire. Sutherlandshire: Ferry-house E. of Loch Erboll.

Ulster.—Antrim: Knockleyd, very rare. Down: Mourne mountains.

Connaught.—Galway: Ma'am Turc.

Leinster.—King's Co.: Tullamore.

Munster.—Kerry: Mucruss, Killarney.

The distribution of this Fern in Europe is general. It is found at North Cape, the extreme northern point of Europe; in Lapland, Sweden, and Russia; in Germany and Hungary; in Dalmatia, Transylvania, and Croatia; in Great Britain; in France, Holland, and Switzerland; in Italy, Spain and Gibraltar. Siberia and Kamtschatka, in Asia, produce it; and it has been recorded from Africa. In the New World it is as widely dispersed as in Europe, occurring in Labrador and Greenland on the north-eastern side; at Sitka, and about Awatschka Bay, the Rocky Mountains, and the Columbia River, on the north-western side; as well as throughout the United States, and in Newfoundland.

This species is a moisture-loving plant, although, as in most other instances, the moisture must not be stagnant. It is also peculiarly a shade-loving Fern; for, though very hardy, and capable of existing under considerable exposure, yet the delicate fronds are damaged and disfigured unless both shade and shelter of some kind is afforded it. It is a good plant for a shady out-door rockery, and also grows readily in pots. Its distinctness of character, and the lovely and refreshing tint of green which it assumes when luxuriating in shade and moisture, no less then its dwarfish size and compact habit, render it one of the most useful of rock Ferns. An admixture of fibry peat and leaf-mould, in the proportions of two-thirds of the former to one-third of the latter, freely mixed with sand, and rubbly sandstone or potsherds, forms a good compost for it. This compost will be suitable either for the open rockery, or for pot culture. When grown in pots, it is best grown in those which are broad and shallow, and must be provided with abundant drainage, as it soon perishes if water stagnates about the roots. The mode of draining a pot or pan efficiently, is to place a thick layer of porous stone or brick or potsherds, broken up to the size of large nuts or walnuts over the bottom; on this a thin layer of the same material of the size of peas, all the fine dusty portions being removed to mix up with the soil; over these hard materials a thin layer of moss is to be spread, to prevent the finer particles of the compost from falling down among the rubble and filling up the vacant spaces. The compost, formed of turfy ingredients, should never be sifted, but the lumpy portions broken up by hand to the sizes of nuts and walnuts, some of these coarser portions of the mass being always placed next the moss. Three or four inches of soil will be sufficient for the plants, as their caudices rather spread near the surface than penetrate. In planting, these caudices should be fixed firmly an inch or so below the surface, which is to be finished of with some of the finer parts of the compost, so that it may be left neat and level. After planting, a good watering through the rose of the watering pot is desirable. When in a healthy vigorous state, the caudices creep about rapidly in all directions. When planted out in the rockery, it should also have perfect drainage, in order that the roots may be freely supplied with water as circumstances require.

This is one of the most interesting of the dwarf annual-fronded British Ferns, for a glass case. The only objection to it is to be found in its deciduous habit, in consequence of which its place in winter becomes a blank. The beautiful tint of its fronds, when they do appear in spring, is however perhaps an ample compensation for this defect in its ornamentative capacities. It increases with facility by division of the caudex.

THE LIMESTONE POLYPODY.

POLYPODIUM ROBERTIANUM.

P. fronds erect, rigid, glandulose, pentangular deltoid, subternate; lower branches (or pinnæ) bipinnate, stalked, their pinnulets or lobulets oblong obtuse, crenate or nearly entire; stipes glandulose. [Plate XII.]

POLYPODIUM ROBERTIANUM, *Hoffmann, Deutschl. Fl.* ii. in addenda to p. 10 (1795). *Fries, Sum. Veg.* 82. *Koch, Synops.* 2 ed. 974. *Nyman, Syll. Fl. Europ.* 430. *Moore, Handb. Brit. Ferns*, 3 ed. 66; *Id., Ferns of Gt. Brit. and Ireland, Nature Printed*, t. 6. *Babington, Man. Brit. Bot.* 4 ed. 420. *Lowe, Nat. Hist. Ferns*, i. t. 28.

POLYPODIUM CALCAREUM, *Smith, Fl. Brit.* 1117 (1804); *Id., Eng. Bot.* xxii. t. 1525; *Id., Eng. Fl.* 2 ed. iv. 270. *Deakin, Florigr. Brit.* iv. 43, fig. 1582. *Hooker & Arnott, Brit. Fl.* 7 ed. 582. *Newman, Hist. Brit. Ferns*, 2 ed. 131. *Sowerby, Ferns of Gt. Brit.* 12, t. 4. *Willdenow, Sp. Plant.* v. 210. *Sprengel, Syst. Veg.* iv. 60. *Presl, Tent. Pter.* 180.

POLYPODIUM DRYOPTERIS, VAR., *Bolton, Fil. Brit.* 53, t. 1. *Bentham, Handb. Brit. Fl.* 627.

POLYPODIUM DRYOPTERIS, β. MINUS, *De Candolle, Fl. Franç.* ii. 565.

POLYPODIUM DRYOPTERIS, β. ROBERTIANUM, *Ruprecht, Dist. Crypt. Vasc. Ross.* 52. *Ledebour, Fl. Ross.* iv. 509.

POLYPODIUM DRYOPTERIS, v. CALCAREUM, *Gray, Man. Bot. North. U. States*, 590.

NEPHRODIUM DRYOPTERIS, *Michaux, Fl. Bor. Amer.* ii. 270.

LASTREA CALCAREA, *Bory, Dict. Class. d'Hist. Nat.* ix. 233. *Newman, Nat. Alm.* 1844, 17.

LASTREA ROBERTIANA, *Newman, Hist. Brit. Ferns*, 2 ed. 13.

PHEGOPTERIS CALCAREA, *Fée, Gen. Fil.* 243. *J. Smith, Cat. Cult. Ferns*, 17.

GYMNOCARPIUM ROBERTIANUM, *Newman, Phytol.* iv. 371; *Id., Appendix*, 1851, xxiv.; *Id., Hist. Brit. Ferns*, 3 ed. 63.

Caudex creeping extensively, branched, thicker than a straw, dark-brown, scaly. *Scales* pale-brown, semitransparent, lanceolate. *Fibres* dark-brown, branched, clothed with a brighter brown pubescence.

Vernation circinate; the pinnules rolled up separately into little globules, the pinnæ then rolled each separate inwards towards the main rachis, which is next itself coiled up.

Stipes longer than the frond, often twice as long, stoutish, succulent when young, becoming stiff and erect, abundantly scaly about the base, and with a few scattered deciduous scales upwards when young, minutely glandular, pale watery-green, dulled by the glandulosity of the surface; lateral to the caudex, adherent, distinct. *Rachis* glandulose, the part forming a stalk to the lower pinnæ much shorter and distinctly smaller than that between the first and second pairs of pinnæ.

Fronds six to eighteen inches in height, including the stipes, which is usually more than half, sometimes two-thirds at least, of the length: erect, of a firm herbaceous texture, deep dull grayish-green, glandulose, elongately deltoid-pentangular, the pentagonal outline, however, less manifest than in *P. Dryopteris*, in consequence of the less comparative length of the stalks of the lower pinnæ. The fronds are not truly ternate, though the larger size of the lower pinnæ gives them a subternate appearance; they are bipinnate, with the lowest pair of pinnæ subbipinnate or sometimes bipinnate on the posterior side, which is the most developed.

Pinnæ variable, opposite below, the lower largest pair sometimes each six inches long, obliquely triangular, stalked, often bipinnate; the next pair stalked or sessile, pinnato-pinnatifid; the upper ones all sessile, pinnate or pinnatifid, becoming gradually less divided towards the apex. *Pinnules* of the lower pair larger on the posterior side, those of the other pinnæ nearly equal; those of each succeeding pair resembling the smaller ones of the pair next below them. *Pinnulets* or *lobulets* oblong, obtuse, entire or crenated.

Venation of the lower posterior pinnules consisting of a stout *costa* or midvein, with a flexuose *vein* running up the centre of each lobulet; this is alternately branched, the *venules* extending to the margin, simple, or very commonly forked, the venule if simple, and the anterior *veinlet* if divided, bearing a sorus near to the margin. Or, the vein extending up the lobulet may be regarded as a midvein; its branches, sometimes simple and soriferous, as veins; and the branches of these, of which the anterior is mostly fertile, as venules.

Fructification on the back of the frond, scattered over its whole surface. *Sori* small, circular, consisting of numerous crowded spore-cases, entirely without indusia, arranged in a linear submarginal series along each side of the lobulets; or about the sinus, in a series between the midrib and margin, when the lobules are but slightly developed; often more or less confluent. *Spore-cases* pale-brown, roundish-obovate, small, numerous. *Spores* ovate, or oblong, somewhat granulate.

Duration. The caudex is perennial. The fronds are annual, the earlier ones growing up about May, and the latest perishing in autumn.

The chief differences between *P. Robertianum* and *P. Dryopteris* have been already pointed out in our remarks on the latter species. The most important of these is the pinnate rather than ternate plan of division of its fronds. This combined with the distinctive features afforded by its stouter, more erect, and more rigid habit, the glandulosity of its entire surface, and its constancy both in the wild and cultivated state, leaves no reasonable ground to doubt its permanent distinctness from its near ally.

We advisedly retain this species, together with *P. Dryopteris* and *P. Phegopteris* in the genus *Polypodium,* from a conviction that it is mere folly to multiply genera on grounds so slight as are depended on for distinguishing them in this instance. Distinctly and unmistakeably characterised among the annulate Ferns by free veins and round naked sori, the genus *Polypodium,* thereby relieved of a host of species having reticulated veins, is perfectly intelligible, and though extensive is not unwieldy. Mr. Newman would, however, separate from *Polypodium,* under the name of *Gymnocarpium,* the three plants above referred to; and so far as any intelligible characters have been assigned to it, this group would be distinguished by having a slender black underground caudex—a feature which is assuredly not of generic value. Presl had indeed at a much earlier date, as we have already remarked, proposed a nearly corresponding group as a section of *Polypodium;* and M. Fée had adopted this group under the name of *Phegopteris* as a genus, in his admirable *Genera Filicum,* distinguishing it by a character which would be of far more importance than the nature of the stem or caudex, if constant, namely, that of having medial fructification, the receptacle of the sori being placed below the apex of the vein. Unfortunately, however, in this very genus, there are species which produce, at the same time, both medial and terminal sori, so that the character is not distinctive. The species referred to this group possess, however, in common, a peculiarity of some importance, their fronds being adherent to, not articulated with, the creeping caudex.

This species occurs we believe exclusively on exposed rocky limestone tracts. It is found plentifully in some parts of England, chiefly in the western, central, and northern districts, from Somersetshire in the south-west, to Durham in the north; in the former district descending to about 250 feet above the sea, and in the north ascending to 900 feet or upwards. It also occurs plentifully in some parts of Wales. The calcareous hills of Gloucestershire seem, however, to be its head-quarters. It is not known to occur in Ireland, in Scotland with the adjacent Isles, or in the eastern counties of England. Its distribution is recorded as follows:—

Peninsula.—Somersetshire: Bath; Cheddar Cliffs; Mendip Hills; Friary Wood; Hinton Abbey.

Channel.—Wiltshire: Box quarries; Corsham, *Dr. Alexander.*

Thames.—Oxfordshire.

Severn.—Gloucestershire: Besborough Common, *W. H. Purchas;* Rocks by the Wye, near Symond's Yat; New Weir; Lydbrook, Forest of Dean; Cleeve Clouds and Windlass Hill, near Cheltenham; Postlip Hill, and elsewhere on the Cotswolds; Cirencester, *J. Buckman;* English Bicknor, *A. T. Wilmot;* Leigh Wood, near Bristol. Herefordshire: Colwall, near Whitchurch. Worcestershire. Staffordshire.

Trent.—Derbyshire: Matlock; Wirksworth; Roadside under the Lover's Leap, Buxton; Bakewell, *T. Butler;* Dovedale.

Mersey.—Lancashire: Lancaster; Sheddin Clough, near Burnley; Broadbank.

Humber.—Yorkshire: Ingleborough; near Settle; Anster Rocks; Arncliffe; Gordale; Ravenscar, Waldenhead, *J. Ward;* near Bradford, *J. T. Newboult;* near Sheffield.

Tyne.—Durham: Falcon Clints, *T. Simpson.*

Lakes.—Cumberland: Newbiggin Wood; Gelt Quarries; Baron Heath; Scale Force, *J. Robson.* Westmorland: Arnside Knot; Hutton Roof Crags; Farlton Knot; Caskill Kirk.

S. Wales.—Glamorganshire: Merthyr-Tydvil. Brecknockshire: River Clydach, near Llanelly, *T. H. Thomas.*

N. Wales.—Denbighshire: Llanferris; near Ruthin, *T. Pritchard.* ? Carnarvonshire: Cwm-Idwal.

The plant appears to be met with over a considerable part of Europe, as, for example, in Norway, Sweden, and Russia, in England, in France, Belgium, and Switzerland, in various parts of Germany, in Hungary, and in Dalmatia. In Asia, it has been gathered by Drs. Hooker and Thomson, in the Himalaya mountains, at an elevation of 5-8000 feet. In North America it occurs both in the United States and in Canada.

This is a hardy plant under cultivation, provided its roots are well drained, and the soil in which it is planted is kept rather drier than is usual with Ferns, particularly in winter. This latter point may be effected, both by withholding excess of water, and by adding to the compost some porous materials, among which limestone, soft sandstone, or old mortar are the most suitable. This species bears a moderate degree of exposure to sun, better than the majority of Ferns. Its habit of growth is the same as that of *P. Dryopteris*, the caudex creeping out in all directions. Hence, when grown in pots, it is like that species, best planted in those which are wide across the mouth; and as neither of these plants root deeply, pots of shallow form, or pans, of which some ornamented patterns are manufactured especially for fern culture, are preferable for them. In the out-door rockery, where the Limestone Polypody succeeds well under favourable conditions, some especial provision must be made for drainage in the spots where it is planted. The creeping caudices are sometimes apt to perish in winter, if they have not been tolerably well ripened, and are not kept from anything like excess of moisture. These caudices afford a ready means of propagation.

GENUS II: ALLOSORUS, *Bernhardi.*

GEN. CHAR.—**Sori** spuriously-indusiate, rotundate, covered by the revolute subherbaceous margin of the pinnules, at length confluent into a transverse line (parallel to the margin), often becoming effuse; the **receptacles** punctiform. **Veins** in the fertile fronds simple or forked, from a central costa; in the more divided sterile fronds simple or forked in the ultimate segments; **venules** free.

Fronds dimorphous, dwarf, herbaceous, bi-tri-pinnate; the fertile contracted, *i. e.*, with revolute siliculiform pinnules.

Caudex short, decumbent.

Few, if any, of the Ferns which are indigenous to Great Britain, have given rise to such conflicting opinions as this, as to the genus to which it belongs. Linnæus, and the older botanists, referred it to *Osmunda* and *Onoclea*; Villars to *Acrostichum;* while of the other names which have been applied to it, all apparently under the impression of its being a pteroid Fern, the *Allosorus* of Bernhardi claims priority, and we adopt it with some limitations.

Allosorus, as here restricted, is a small genus of three or four dwarf elegant parsley-like Ferns, widely scattered over the globe. Perhaps it should be united to *Cryptogramma*, with which it was doubtfully associated by the author of the latter genus, the only material difference between them being that *Allosorus* has punctiform receptacles, whilst in *Cryptogramma* the receptacles are linear and oblique. In habit and aspect they are quite alike. We follow Mettenius and others, in keeping them distinct, on account of the difference in the receptacle, to which we attach considerable importance. In the typical species of *Cryptogramma*, the sori form short lines along a portion of the veins, after the gymnogrammoid type, and these lines being parallel, and near together, unite laterally as they become effused, and so form a broad linear mass transverse to the veins. In *Allosorus*, the sori instead of being elongated are punctiform, but they become laterally confluent in the same way as in *Cryptogramma*, and in some states of the plant a tendency to elongate is perhaps also to be observed. The two groups are undoubtedly very closely related, and we have regarded them as constituting the salient points, at which the genera having linear and punctiform sori, touch each other.

Sir W. J. Hooker, in his *Species Filicum* (ii. 127), has not only united our *Allosorus* with *Cryptogramma* under the latter name, but also several species of the two supposed genera into one, represented by *A. crispus*. "If indeed," he observes, "there was a manifest difference in the sori, so as to constitute different genera, between *C. crispa* and *C. acrostichoides* and *Brunoniana*, as Presl and lately Mettenius maintain is the case, the first could on no account be united with the two latter, but I think I may appeal to the magnified representations of the sori of *C. crispa* as given in our *Genera Filicum*, and in Fée's *Genera Filicum*, and of those of the other two kinds in the *Icones Filicum*, in support of my views that there is no available distinction." "When," he continues, "an old plant is found in a part of the world very distant from its previously known locality, one is apt to look upon it as something new; and, as is the case with the Cedar of Lebanon and the Cedar of the Himalaya, it is very difficult to remove the impression once made on the mind, although no tangible character to distinguish them can be detected." We think the figures of *A. crispus* referred to are defective.

The genus *Allosorus* was originally proposed by Bernhardi for a group of very varied Ferns, having no combining character which can now be considered as important. The name has since his day been variously applied, but by the generality of pteridologists it has been assigned to a group of doubtful plants, oscillating between *Pteris* and *Cheilanthes* according to the particular views of authors, but nevertheless quite devoid of any satisfactory character, to distinguish them from these genera; some of the species having punctiform receptacles, and belonging truly to *Cheilanthes*, notwithstanding their having a continuous indusioid margin, and others being truly pteroid, having the sori attached to a continuous marginal receptacle. There being, consequently, no place for a cheilanthoid genus *Allosorus*, we prefer to follow those who retain the name for the Parsley Fern, which was one of the original species of Bernhardi.

There has not only been a difference of opinion as to the generic name of this Fern, but also as to its affinity. By some it has been considered as of pteroid structure, and this view may be true of some of the species already mentioned as having been referred to *Allosorus* since Bernhardi's time, but does not well apply to our present plant. *Pteris* has a continuous marginal receptacle, and pteroid plants should have the same, but there is nothing of the kind in the *Allosorus crispus*. Its receptacles are punctiform, as in *Polypodium*, and when, as in the allied *Cryptogramma*, there is any deviation from this structure, it is not towards the production of a transverse marginal receptacle, but the opposite—an oblong sorus parallel to the venation. These receptacles, if the construction of the sori is of any value, have an undoubted affinity with *Platyloma*, which we think is properly considered a peculiar type of development; and we have no difficulty in coming to the conclusion that whether or not truly distinct from *Cryptogramma*, and consequently whether belonging to the *Platylomeæ* or the *Polypodieæ*, the plants referred to these two genera indicate the points at which the two groups coalesce, and by which the punctiform *Polypodieæ* become connected through the *Platylomeæ*, with the line-fruited *Gymnogrammeæ*, or *Grammitideæ* as they are sometimes called. The plant is, in fact, polypodio-grammitoid, the sori being round or oblong, and distinct. There is nothing whatever in *Allosorus* but the reflexed indusioid margin, which resembles the structure of *Pteris;* and although this may produce a considerable degree of outward similarity, yet the punctiform receptacles and the non-indusiate sori of *Allosorus* at once distinguish it.

The name of the genus is derived from the Greek, *allos*, various, and *sorus*, a heap.

BRITISH SPECIES.

A. crispus; a dwarf perennial, with dimorphous twice or thrice pinnate fronds.

THE MOUNTAIN PARSLEY FERN, or ROCK BRAKES.

ALLOSORUS CRISPUS.

A. fronds of two kinds, ovate-deltoid, bi-tri-pinnate; ultimate divisions of the sterile fronds obovate wedge-shaped, often bifid; those of the taller fertile fronds linear or oblong, their margins recurved over the roundish sori. [Plate XIII.]

ALLOSORUS CRISPUS, *Bernhardi, Schrad. N. Journ. Bot.* 1806, i., pt. ii. 5, 36. *Babington, Man. Brit. Bot.* 4 ed. 410. *Deakin, Florigr. Brit.* iv. 47, fig. 1585. *Newman, Hist. Brit. Ferns,* 3 ed. 35. *Moore, Handb. Brit. Ferns,* 3 ed. 70; *Id., Ferns of Gt. Brit. and Ireland, Nature Printed,* t. 8 (excl. syn. Ruprecht and Gmelin); *Id., Ind. Fil.* 44. *Sowerby, Ferns of Gt. Brit.* 69, t. 39. *Sprengel, Syst. Veg.* iv. 65. *Presl, Tent. Pter.* 152. *Koch, Synops.* 2 ed. 985. *Bentham, Handb. Brit. Fl.* 627. *Lowe, Nat. Hist. Ferns,* iii. t. 34. *Nyman, Syll. Fl. Europ.* 434.

OSMUNDA CRISPA, *Linnæus, Sp. Plant.* 1522. *Bolton, Fil. Brit.* 10, t. 7. *Hudson, Fl. Ang.* 450. *Flora Danica,* t. 496. *Savigny, Lam. Enc. Bot.* iv. 657.

OSMUNDA RUPESTRIS, *Salisbury, Prod.* 402.

PTERIS CRISPA, *Linnæus MS. Smith, Fl. Brit.* 1137; *Id., Eng. Bot.* xvii. t. 1160; *Id., Eng. Fl.* 2 ed. iv. 306. *Schkuhr, Krypt. Gew.* 90, t. 98. *Willdenow, Sp. Plant.* v. 395 (excl. syn. Gmelin).

PTERIS TENUIFOLIA, *Lamarck, Fl. Franç.* i. 13.

ACROSTICHUM CRISPUM, *Villars, Hist. des Pl. Dauph.* iii. 838.

ONOCLEA CRISPA, *Hoffmann, Deutschl. Fl.* ii. 11.

CRYPTOGRAMMA CRISPA, *R. Brown, App. Frankl. Narr. of Journ. to Polar Sea,* 754, 767. *Hooker, Gen. Fil.* t. 115 B (sori too long); *Id., Sp. Fil.* ii. 128 (European form, excl. syn. Gmelin, Turczaninow, and Ruprecht). *Hooker & Arnott, Brit. Fl.* 7 ed. 590. *Mackay, Fl. Hib.* 343.

PHOROLOBUS CRISPUS, *Desvaux, Ann. Soc. Linn. de Paris,* vi. 291. *Fée, Gen. Fil.* 131, t. 7 D.

BLECHNUM CRISPUM, *Hartmann, Fl. Scand.* 3 ed. 255.

RIEDLEA CRISPA, *Mirbel.*

STEGANIA ONOCLEOIDES, *Gray, Nat. Arr. Brit. Pl.* ii. 16.

STEGANIA CRISPA, *R. Brown, Prod. Fl. Nov. Holl.* 152, in obs.

STRUTHIOPTERIS CRISPA, *Wallroth, Bluff et Fingerhuth, Comp. Fl. Germ.* iii. 27.

Caudex small, short, tufted, erect or decumbent, scaly. *Scales* membranous, pale brown, subulate. *Fibres* numerous, branched, dark brown, wiry, and slightly covered with small hair-like scales.

Vernation circinate.

Stipes as long as, or usually longer than the frond, pale green, slender, smooth, with a few scattered scales near the base; lateral and adherent to the caudex. *Rachis* smooth.

Fronds from four to twelve inches high, including the stipes, herbaceous, of a lively green, terminal on the caudex, triangular or ovate-triangular in outline, of two forms, and hence described as dimorphous. *Sterile fronds* leafy, usually about as long as the stipes, bi- or tri-pinnate, smooth. *Pinnæ* alternate or sub-opposite, triangular-ovate, spreading, the lower ones largest. *Pinnules* alternate, ovate, largest on the lower side of the pinnæ, pinnate or pinnatifid, the pinnulets or lobes ovate, or obovate-cuneate; the latter or smaller ones, cut into linear acute teeth, and the former into cuneate-linear bifid lobules, having acute incurved teeth. The ultimate divisions are, however, variable in form, being sometimes oblong-oval, with sinuously shallow-toothed margins, this form of development apparently representing fertile fronds, whose fructiferous growth has become arrested and abortive. *Fertile fronds* contracted, usually about one half as long as their stipes, tripinnate or in some cases quadripinnate in the basal portions of the lower pinnæ. *Pinnæ* alternate or sub-opposite, ovate, spreading, the lower ones largest. *Pinnules* alternate, ovate in outline, bipinnate or pinnato-pinnatifid in the lower pinnæ, pinnate only above. All the ultimate divisions are stalked, obtuse, and linear-oblong from the involution of the margins, which are pale-coloured, crenated, and indusioid.

Venation of the barren fronds consisting of a slender *costa* extending along each pinnule, and casting off a vein into each of its lobes or pinnulets, this again becoming alternately branched, so that a *venule* or *veinlet* runs along the centre nearly to the point of each segment —simple where the segment is undivided, and forked where it is bifid, one branch being directed towards every marginal tooth. In the fertile fronds a *costa* or midvein enters each ultimate division, and passes in a sinuous course to its apex, throwing out alternate *veins* which extend nearly to the margin, and are usually simple but sometimes forked and bear a sorus near to their extremity.

Fructification on the back of the frond, and usually occupying the whole under surface. *Sori* small, roundish, situated near the extremity of the venules; at first distinct though contiguous, ultimately becoming laterally confluent and forming a continuous line.

Indusium none, but the margins of the pinnulets, somewhat pallid but not altered in texture, are incurved over the sori. *Spore-cases* small, elliptic-obovate, stalked. *Spores* smooth, roundish, oblong, or bluntly triangular.

Duration. The caudex is perennial; the fronds are annual, springing up in May and June, and perishing in the course of the autumn.

The Mountain Parsley Fern is readily known by its dwarf tufted parsley-like appearance, coupled with the dissimilarity between its much-divided sterile and fertile fronds, of which the former have the segments broad, flat, and leaf-like, and the latter have them involute at the margin, so that they become contracted and somewhat pod-like or siliculiform. These features distinguish it from all others of our native Ferns.

This Fern is met with rather plentifully, though locally, on the mountains of Scotland and those of the northern parts of England, and occurs sparingly in a few scattered stations, in Devonshire, and the districts of the Mersey, the Trent, and the Severn. In Wales it occurs, though not abundantly, in several counties in the north, including the Snowdon district; in South Wales it is more rare. In Ireland it is also rare, being recorded only from three or four counties. It is a mountain plant, preferring rocky situations, and delighting to grow among boulders and loose stones, or on stone walls, where it is protected from excess of moisture. Mr. Watson calls it rupestral and pascual in its habits. It occurs nearly at the sea level in the moors of Lancashire, and in North Wales descends to about 450 feet; while in the West Highlands it ascends to an elevation of upwards of 3000 feet—1150 yards according to Mr. Watson. The following are the recorded habitats:—

Peninsula.—Devonshire: Exmoor near Challacombe, *N. Ward.* Somersetshire: Simmonsbath. These descriptions perhaps refer to one locality.

Severn.—Shropshire: Titterstone Clee Hill. Worcestershire: Herefordshire Beacon, Malvern Hills. ? Staffordshire: Stowe.

Trent.—Derbyshire: Fairfield; Chinley Hill, near Chapel-en-le-Frith. ? Rutland.

Mersey.—Cheshire: Tag's Ness near Macclesfield. Lancashire: Lancaster; Cliviger near Todmorden; Thevely near Burnley; Foedge near Bury.

Humber.—Yorkshire: Settle; Penhill; Saddleworth; Fountain's Fell; Haworth near Halifax; Wensleydale; Cronkley Scar; Ingleborough, etc.

Tyne.—Durham: Falcon Clints, Teesdale; Cocken; Walls near Cronkley Fell. Northumberland: Cheviots above Langley Ford; Crag Lake; Haltwhistle.

Lakes.—Westmoreland: Ambleside; Casterton; Old Hutton; Kendal; Morland; and elsewhere on the hill-sides, abundant. Cumberland: Borrowdale; Ennerdale; Derwentwater; Winlatta, *W. Christy;* Grassmere; Keswick; Skiddaw; Scawfell; Helvellyn; Saddleback; Martindale near Wigton, *W. G. Johnstone,* etc. N. Lancashire: Conistone. Isle of Man, *Dr. Allchin.*

S. Wales.—Glamorganshire: Aberdare. Cardiganshire.

N. Wales.—Denbighshire: Cerig-y-Druidion; Ruthin, *T. Pritchard.* Merionethshire: Dolgelly; Cader Idris. Montgomeryshire: Breiddin hills. Carnarvonshire: Cwm-Idwal; Clogwyn-du-Yrarddu, Snowdon; Glyder Vawr; Mynidd-Mawr; Llanbaba, *W. Pamplin;* Llanberis; Aber; and elsewhere.

W. Lowlands.—Dumfriesshire: Dumfries; Jardine Hall; George Town; Queensberry hill; Rae hill; hills above Loch Skew; Morton hills; Moffat-dale, *P. Gray.* Kirkcudbrightshire: Sandy hills and Douglass hall, Colvend; Carsethorne, *P. Gray;* Criffel. Ayrshire: Cuff-hill and Beith. Renfrewshire: Neilston Pad, *W. L. Lindsay.*

E. Lowlands.—Roxburghshire: Eildon hills; Winchope, *W. Scott.* Berwickshire: south bank of the Whiteadder. Edinburghshire.

E. Highlands.—Fifeshire: West Lomond Hill; Saline Hill. Perthshire: Summit of Ben Ledi, *Mrs. Macleod;* Ben Lawers; Killin; Dunkeld, *A. Tait;* Glen Tilt; Blair Athol, etc. Forfarshire: Sidlaw hills, *G. Lawson;* Glen Isla, *W. Brand;* Clova mountains. Aberdeenshire: Glen Callater, *W. Christy;* Castleton; Loch-na-gar, *H. M. Balfour.* Inverness-shire: Kingussie, *A.*

Rutherford; stone walls near Dalwhinnie, and on the neighbouring mountains. Morayshire.

W. Highlands.—Western Inverness-shire: Ben Nevis; Gnarrow; Ben Auldor. Argyleshire. Dumbartonshire: Tarbet, Loch Lomond. Arran: Goat Fell, *J. R. Cobb.* Skye: Ben-na-Caillich. Isle of Mull.

N. Highlands.—Ross-shire. Sutherlandshire.

W. Isles.—Harris: Roddal.

Ulster.—Antrim: Carrickfergus. Down: Sleive Bignian; Mourne Mountains.

Leinster.—Louth: Carlingford Mountain.

Munster.—Clare: Black Head, *E. T. Bennett.*

The species is widely dispersed over Europe in Alpine and subalpine situations, occurring to the north, in Lapland, Norway, Sweden, and Denmark; again, in Great Britain and Ireland, in Germany, Hungary, Switzerland and France; and thence extending southwards into Spain and Italy. According to Sibthorp it grows on Mount Olympus in Asia Minor. It is found at Sitka, and Isle Royal in Lake Superior, in the latter habitat assuming a rather more slender form. Kaulfuss reports it from Unalaschka, an island in the North Pacific Ocean; but this plant, which Ruprecht names *Allosorus foveolatus,* is probably rather identical with the *Cryptogramma acrostichoides.* The East Indian *Cryptogramma Brunoniana* approaches near to the European *A. crispus;* indeed, Sir W. J. Hooker without hesitation includes in his *C. crispa,* both *Cryptogramma acrostichoides* and *C. Brunoniana,* the former as the American, the latter as the Indian form. We have already mentioned the difference in the receptacles which induces us to keep them distinct.

This Fern is not difficult of culture under conditions which protect its fronds from the sun, and its roots and caudex from stagnant or accumulated moisture. It is very beautiful when in a thriving state, being small, bright green, and elegantly divided. Its size adapts it thoroughly for a Wardian case, and it is equally appropriate for culture in a cold frame, or Fern house; indeed we have somewhere seen it mentioned as the pet pit pot-Fern. In its wild state it covers large patches on the sides of rocky mountains, and, as well observed by Mr. Francis, in his book on British Ferns, adds a bright gleam of verdure and of beauty to its romantic but barren dwelling-place; nor does it refuse to give out its ray of cheerfulness and loveliness when transferred to the artificial rockery or Fern-house. On the contrary, in free well-drained soil, and in a cold shady frame, it grows remarkably well, but it is essential that it should be guarded against damp whilst dormant in winter: indeed, at no time should moisture become stagnant about it. The proper soil for the roots consists of turfy peat, freely intermixed with silver sand and with pieces of broken bricks or potsherds, and more sparingly with sandy loam. It is a stone-loving plant, and hence is well suited for artificial rockeries. It may be increased by division, but it is safer not too often to disturb a thriving plant for this purpose.

The plant is rather apt to die off in winter, especially if kept too damp, or not sufficiently drained, so that recourse to its native haunts becomes perhaps often necessary for a supply of plants. It may be useful to hint that, in the case of this, and other Ferns which naturally occur among rocks, and are consequently somewhat difficult to remove and establish, it is far better to select the younger and smaller plants for the purpose of removal, than the larger and older masses which are apt to tempt the collector's hand.

Genus III: GYMNOGRAMMA, *Desvaux*.

Gen. Char.—**Sori** non-indusiate, linear, sometimes elongated, simple or forked *i. e.*, bi-partite, oblique, often at length confluent; the **receptacles** elongate above or continued below the forks of the veins. **Veins** simple or forked from a central costa, or the costa sometimes indistinct; **venules** free.

Fronds lobed pinnate or bi-pinnate, herbaceous or sub-membranaceous, often farinosely ceraceous, sometimes lanate beneath.

Caudex short, erect, sometimes annual.

This is a tropical genus, of considerable extent, and embracing species of very diverse aspect. Indeed so varied are the appearances presented by the plants commonly referred here, that it has been proposed to distribute them into several minor genera, and of these, that which has received the solitary species found within the politico-geographical boundary of Great Britain, has been called *Anogramma*. M. Fée alone proposes or adopts seven of these new genera for the species which we include in *Gymnogramma*, depending for his distinctions chiefly—on the length of the sori: whether occupying nearly the whole length of the veins, or confined within more determinate limits; on the nature of the frond-surface: whether smooth, or covered with hair, or a coloured waxy powder; on the presence or absence of hairs among the spore-cases; or on the simple or divided character of the fronds. There are, however, no sufficiently definite limits to the groups thus indicated to admit of their adoption, even as sectional groups.

Among the species which have often been referred to *Gymnogramma*, there are some having the veins netted, which we think are properly separated, to form the genus *Dictyogramma;* and another plant, having the lines of spore-cases on the free veins confined to a zone near the centre of the kidney-shaped fronds, and so near together as to become laterally confluent into a broad horse-shoe-shaped band, forms the genus *Pterozonium*, which may be considered distinct.

The characteristic feature of *Gymnogramma* is the *forking* of the linear sometimes much elongated sori, which forking, though not occurring in the case of every sorus, does occur more or less frequently over every frond. The spore-cases are ranged in lines along the back of the veins, sometimes in a very scattered manner, sometimes more crowded, and these lines of spore-cases being continued downwards more or less frequently past the point where the veins branch, become like them forked, which is the leading characteristic of the genus. A linear dorsal naked sorus, constantly simple, *i. e.* without this forking, indicates the genus *Grammitis*, which is closely allied to *Gymnogramma*.

Our native Channel Island species, *Gymnogramma leptophylla*, belongs to the § *Pleurosorus*, characterised by having short or shortish lines of spore-cases, the fronds being smooth or hairy. This group contains another annual species, from the West Indies, *Gymnogramma chærophylla*, as well as several others not annual, of a different aspect. The remaining sections are:—§ *Ceropteris*, in which the sori are much as in the § *Pleurosorus*, but the fronds are farinose-ceraceous beneath, as represented by *Gymnogramma chrysophylla;* § *Eriosorus*, with sori as in the last, but the fronds lanate beneath, as in *Gymnogramma lanata;* and § *Neurogramma*, the most distinct of any, in which the sori form long parallel-forked lines, often approximate, and closely placed over all the under surface of the fertile parts; this group, moreover, containing some species which have smooth and others which have hairy fronds, is represented by the South American *Gymnogramma tomentosa*, and the Eastern *Gymnogramma javanica*.

The name is derived from the Greek *gymnos*, naked, and *gramme*, a line.

BRITISH SPECIES.

G. leptophylla a small, fragile, annual, with twice or thrice pinnate fronds.

THE SMALL-LEAVED GYMNOGRAM.

GYMNOGRAMMA LEPTOPHYLLA.

G. fronds oblong ovate, bi-tri-pinnate, glabrous, fragile; pinnæ ovate; pinnules or pinnulets ovate-cuneate, usually three-lobed, the lobes blunt and bidentate. [Plate XIV.]

GYMNOGRAMMA LEPTOPHYLLA, *Desvaux, Berlin Mag.* v. 305; *Id., Journ. de Bot.* i. 26; *Id., Ann. de Soc. Linn. de Paris,* vi. 215. *Moore, Handb. Brit. Ferns,* 3 ed. 74; *Id., Ferns of Gt. Brit. and Ireland, Nature Printed,* t. 43 B. *Newman, Hist. Brit. Ferns,* 3 ed. 12. *Hooker & Arnott, Brit. Fl.* 7 ed. 580. *Babington, Man. Brit. Bot.* 4 ed. 427. *Sowerby, Ferns of Gt. Brit.* 83, t. 48. *Bentham, Handb. Brit. Fl.* 627. *Lowe, Nat. Hist. Ferns,* i. t. 7. *Sprengel, Syst. Veg.* iv. 40. *Presl, Tent. Pter.* 219. *Hooker fil. Fl. N. Zeal.* ii. 45. *Hooker & Greville, Icon. Fil.* t. 25. *Nyman, Syll. Fl. Europ.* 433.

GYMNOGRAMMA PALLISERENSE, *Colenso, Hb. Hooker.*

GYMNOGRAMMA NOVÆ-ZELANDIÆ, *Colenso, Tasm. Phil. Jour.* ii. 165.

POLYPODIUM LEPTOPHYLLUM, *Linnæus, Sp. Plant.* 1553; *Schkuhr, Krypt. Gew.* t. 26.

ACROSTICHUM LEPTOPHYLLUM, *De Candolle, Fl. Franç.* ii. 565.

GRAMMITIS LEPTOPHYLLA, *Swartz, Syn. Fil.* 23, t. 218, 1, fig. 6. *Willdenow, Sp. Plant.* v. 143.

ANOGRAMMA LEPTOPHYLLA, *Link, Fil. Sp.* 137. *Fée, Gen. Fil.* 184, t. 19 A, fig. 1.

ASPLENIUM LEPTOPHYLLUM, *Cavanilles, Anal. de Cienc. Nat.* v. 155, t. 41, fig. 3.—mala, fide Swartz; *Desvaux, Ann. de Soc. Linn. de Paris,* vi. 277.

HEMIONITIS LEPTOPHYLLA, *Lagasca, Gen. et Sp.* 33.

OSMUNDA LEPTOPHYLLA, *Savigny, Lam. Enc. Bot.* iv. 657.

DICRANODIUM, *Newman, Hist. Brit. Ferns,* 3 ed. 13.

Caudex small, subglobose, with a few fine scattered hair-scales in the younger stages. *Fibres* few, brown, pilose.

Vernation circinate.

Stipes as long as, or sometimes longer than, the fronds, smooth and shining, dark chestnut-brown, paler upwards, rather stout in the most perfect fronds; terminal, and adherent to the caudex.

Fronds about six or eight in number, variable in size and form, delicately membranaceous, fragile, pale yellowish green, very slightly hairy when young, quite smooth afterwards. The first frond developed from the prothallus or marchantiform scale, is small, about half an inch long, flabelliform, three-lobed, each of the segments again dichotomously lobed, the lobules blunt and bifid. The next frond acquires an oblong ovate outline, and the three lobes are so far separated as to form three pinnæ, which are divided on the same dichotomous plan as the former; in one such example now before us, which is five-eighths of an inch long, the pinnæ are each twice dichotomously lobed, and each ultimate lobe has its sides nearly parallel, and its apex blunt and two-cleft. Two or three fronds of this pinnate character, each successive one larger and more divided than the preceding, and all broader and more leafy in character than the subsequent ones, are produced during the adolescent state of the plants. After this stage has been passed, the fronds acquire height and become more compoundly divided, and in two, three, or four stages, according to the vigour of the individual plant, reach to their full development. The intermediate fronds are from one-and-a-half inch to three inches high, and are distinctly bipinnate, and generally fertile. The fully developed or mature fronds are from three to six or eight inches high, and grow erect; these are oblong ovate, bi- or tri-pinnate, and fertile throughout. *Pinnæ* ovate triangular, alternate. *Pinnules* ovate wedge-shaped, about three-lobed, the lobes obovate, and notched at the apex; they are scarcely stalked, their base tapering down to a narrow and slightly decurrent point of attachment. Specimens of vigorous growth become tripinnate, by the more complete separation of the lobes of the pinnules.

Venation of the ordinary pinnules consisting of a *costa* which forms by dichotomy a branch, *i. e. vein*, at the base of each lobe; this vein becomes again branched in the same dichotomous manner near the centre of the lobe, its two *venules* being directed, one towards each of the two apical teeth, and terminating within the margin. Occasionally the lobe is not toothed, and the vein is simple.

Fructification occupying the whole back of the frond, without covers. *Sori* linear, forked, occupying nearly the entire length of the venules, and a portion of the vein below the dichotomy, hence forked, that is, diverging in two lines from near the base of the pinnæ along the narrow lobes nearly to their apex, at first distinct, but eventually becoming confluent into one mass. When the vein is simple the sorus is simply linear. *Spore-cases* nearly globose. *Spores* roundish or bluntly angular, faintly striato-punctate, dark brown-purple.

Duration. The caudex is annual, and the development of the plant consequently rapid. In the wild state we learn that the prothallus is developed in the damp late autumnal months, being perfectly formed in November; by January three or four fronds have been produced; in April or May the growth is mature; and by August the plants have perished. Sometimes in cultivation the perfect fronds are not produced till the second year.

This Fern clearly belongs to the genus *Gymnogramma*, which is distinguished from *Grammitis* by the greater length of the linear sori, and their more or less frequently forked condition. This group, though itself not too distinct from *Grammitis*, which has simple oblong sori, some modern botanists have separated into several genera, and one of these, *Anogramma*, was proposed expressly for the reception of this species. Beyond certain peculiarities of habit and aspect there is, however, nothing to separate generically any of the free-veined *Gymnogrammas*, and such distinctions as these alone are insufficient. Mr. Newman calls this plant the Annual Maidenhair.

No other British Fern approaches at all nearly to the Small-leaved Gymnogram, either in aspect, or in botanical characters.

The habitat which brings this species within the British Flora, namely, the Island of Jersey, is to be understood as British rather politically than geographically. In that island the plant occurs in several ascertained localities, principally in the neighbourhood of St. Laurence, of St. Aubin, and St. Haule. It is said to have been first found by "a lady," in 1852; but the earliest public notice of the discovery, as far as we know, occurs in the early part of 1853, in one of the horticultural periodicals,* where, under the signature of J. M., we read:—"Your assurance that my Fern is *Gymnogramma leptophylla*, an entirely new fern to the Flora of

* *Gardeners' Chronicle*, Jan. 29, 1853, p. 69.

Great Britain, is highly gratifying. This morning I examined the place where it was gathered last year, and find that it is coming up plentifully again. It is growing in a clay soil, on a bank at the foot of a hill, and is much overshadowed with ivy and larger Ferns; the *Asplenium lanceolatum*, too, grows plentifully all around it, and the bank in that part is covered with a small round lichen [perhaps *Marchantia*]. The situation is very damp and much sheltered, and the Fern is scattered over a surface of two or three yards; but I can find no trace of it on any other part of the bank, and I have never met with it in any other part of the island. The place where it grows is unfrequented, and I do not think it is possible it should be anything but wild." Subsequently * Mr. Newman also recorded its discovery, and mentions one spot near St. Laurence where it grows plentifully for a considerable distance along a hedge bank, extending as far as the bank is exposed, but ceasing exactly where the lane is shaded by trees. There can thus be no doubt the species is indigenous in the Channel Islands. Mr. Ward, who visited its localities in 1853, informs us that he found it growing on the exposed banks of lanes with a south-western aspect, protected from the sun by the surrounding vegetation which clothes the banks, and fed by the constant oozing of water, which renders the soil sufficiently moist for the growth of Liverwort and Mosses. According to another report,† it was found in 1852, by Miss Veitch, "in a stone dyke on the high road leading from Braemar to Ballater, nearly opposite Invercauld House," in Aberdeenshire, but as no further evidence of its existence there has been forthcoming, and the habitat seems too far north for a tender Fern, this report probably originated in accidental error, and perhaps from the chance intermixture of Scottish and Madeira dried plants.

This delicate species is remarkable on account of its wide dispersion over the world. In Europe it ranges from Jersey, France, and Switzerland, its northern limits, into Germany, extending to Spain, Portugal, and Gibraltar, on the one hand, and Italy: Naples, Sicily, and Sardinia; Corsica, Dalmatia, Greece, and Crete, on the other. In Asia it is found in India: at Mussoorie and in the

* *Phytologist*, iv. 914, March, 1853; and 973, June, 1853. † *Id.* iv. 600.

Neilgherry Mountains; at Ghilan and Lazistan, according to Ruprecht; and in the island of Karek in the Persian Gulf. In Africa, it occurs in Algiers, Morocco, Egypt, Abyssinia, and in the Atlantic Isles—the Canaries, Madeira, Teneriffe, the Azores, and the Cape de Verds, in the northern hemisphere; and at the Cape of Good Hope, in the southern. In America, it is found in Mexico at Vera Cruz. In Australasia, it is found at Victoria and the Swan River, and in Tasmania and New Zealand.

This plant succeeds with very little care from the cultivator, and like its West Indian ally, *Gymnogramma chærophylla,* also an annual, scatters its spores, and becomes, as it were, a weed in congenial situations. Any light sandy soil seems to suit it. That in which it grows naturally in some parts of Jersey, and of which Mr. Ward kindly gave us a portion richly furnished with its spores, is a sandy loam; and scattered on the surface of a flower-pot, filled with similar soil, this earth yielded an abundant crop of plants. The young plants like shade, moisture, and a temperate climate, which conditions will ensure their successful growth. Propagation must either be trusted to the natural scattering of the spores, or a frond or two just arrived at maturity should be preserved and the spores deposited towards autumn in the situations where plants are required. We learn from several cultivators, who have grown the plant in cold situations, that the development has not gone beyond the production of the prothallus until the second year. Our plants, in a warm situation, have been strictly annual, that is, sown in autumn they have matured their fronds early in the following summer. In order to prevent the drying of the surface either before the spores have vegetated or subsequently, it is advisable to keep the soil covered by a bell-glass. The *Gymnogramma* will succeed well by preparing a pot half-filled with drainage, and the rest with sandy loam and lumps of freestone, and then scattering the spores thinly over the surface, which is to be kept enclosed. No transplantation is then necessary, nor need the plants be disturbed unless they vegetate too thickly, when a portion may be carefully thinned away.

Genus IV: POLYSTICHUM (*Roth*), *Schott.*

Gen. Char.—**Sori** indusiate, globose; the **receptacles** medial or rarely terminal on the venules. **Indusium** orbicular, peltate. **Veins** pinnato-furcate or simply forked, from a central costa; **venules** free, the lower anterior one usually, sometimes more, fertile.

Fronds simple pinnate or bi-tri-pinnate, rigid, coriaceous, the margins usually mucronato-serrate.

Caudex short, thick, erect.

This extensive genus is very well marked by technical characters derived from the fructification. The original *Polystichum* of Roth, *Aspidium* of Swartz, and *Tectaria* of Cavanilles, all proposed about the same date, were intended to separate the indusiate group at that time referred to *Polypodium*, from among the typical non-indusiate species. In the disposition of the two former of these names, long since made by Schott and adopted by Presl, by which *Aspidium Lonchitis* was made the type of *Polystichum*, and *Aspidium trifoliatum* of *Aspidium*, we entirely concur; though it may be regretted that either the expressive name proposed by Cavanilles, or the still older *Dryopteris* of Adanson, was not used by Presl, instead of the more modern inexpressive name, *Lastrea* of Bory, for the group of which the old *Aspidium Filix-mas*, and *Aspidium dilatatum* are the types. This latter name, *Lastrea*, having been, however, employed so long ago, both in the arrangements of Presl and J. Smith, on which modern views of classification are mainly based; and the group being so extensive, that the substitution of another generic name would involve multitudinous changes, it is doubtless better now to acquiesce in the nomenclature of Presl, so far as regards the application of the name of *Lastrea* to the free-veined reniform *Aspidieæ*, and that of *Nephrodium* to those having anastomosing veins, than to adopt any other, or attempt a redistribution of the species. Professor Fée, indeed retains the name of *Aspidium* for the *Lastrea* group, on the ground of its containing the larger proportion of the original species; but it seems more consistent to retain the typical plants of Swartz, and no species has so strong a claim to be considered typal as *A. trifoliatum*, to which the name was allotted by Schott. Swartz assigns the characters "Sori subrotundi, sparsi; indusio umbilicato l. dimidiato tecti" to his genus *Aspidium*, thus giving precedence to the umbilicate, *i. e.* peltate indusium. Hence, therefore, *A. trifoliatum*, the first peltate species in his enumeration, may fairly be considered typal, and this has reticulated venation.

The peculiar characteristics of the genus *Polystichum* consist in the punctiform sori being dorsal on the free veins, and covered by circular peltate indusia. It consequently differs from true *Aspidium* as above indicated, in the free instead of reticulated venation; whilst from *Lastrea*, with which it agrees in having free veins, it differs in having peltate instead of reniform or dimidiate indusia. From *Polypodium*, the *Phegopteris* section of which has similarly placed punctiform sori, it is distinguished by the presence of the indusium.

Polystichum is not always readily known from *Polypodium*, on account of the fugacious character of the indusium in some species, for if this is cast off in an early stage, as it frequently is, nothing remains by which to distinguish them. There are some exotic species, not very satisfactorily determined from this cause: being only known from herbarium specimens in which it is uncertain whether the indusium has never been present, or has been cast off while young, the sori being apparently naked from whichever cause.

Polystichum is an extensive genus, consisting for the most part of harsh evergreen spiny-toothed Ferns, scattered from the torrid to the frigid zones, and represented by two or three species in our own country. These species are so variable, and so thoroughly connected by intermediate forms, that it is difficult to come to a conclusion as to their limits which is perfectly satisfactory to botanists generally. Some persons admit three native species, and we adopt this view; others refer all the forms to two species. The series is so perfect from beginning to end, that it appears to us the only alternatives are to receive three species, or one only; and we do not find this latter course anywhere maintained. In the case of British specimens there is little difficulty, if any, in distinguishing three species. It is only when exotic forms are also taken into account, that serious difficulty arises. Even in that case, however, we rather prefer to consider this as one of those numerous instances in which specific limit is not clearly definable, than as an example of specific identity under phases of extreme diversity of character. Indeed, if such comprehensive species are admitted, it will become utterly impossible to define them.

The majority of the species of *Polystichum*, including all those of British origin, belong to the § *Hypopeltis*, which may be considered as the typical group, distinguished by having its pinnæ and pinnules continuous with the rachis. The § *Cyclopeltis*, which we also include in the genus, and which is represented by the West Indian *Aspidium semicordatum* of Swartz, differs only in having the pinnæ articulated with the rachis.

Beyond the foregoing, there is a group of species, represented by the old *Aspidium aristatum*, in which the indusium, which is roundish in outline, has only a shallow lateral notch, from which a furrow or depression extends inwards towards the centre; being attached at this part, and consequently towards the centre, many botanists have regarded the indusium as being peltate, and referred the plants to the genus *Polystichum*. In this view we do not agree. In reality, the indusia in the plants now referred to, differ only in degree from those of the Common Male Fern, being obviously roundish-reniform, so that these species, though polystichoid in habit, must be placed under *Lastrea*.

The English name of Shield Fern was that given to the old genus *Aspidium*, and we therefore adopt it for the present group on account of its being the most typical of the two genera under which the English species of *Aspidium* are placed. The peltate involucre sufficiently justifies this adoption.

The generic name is derived from the Greek *polys*, many, and *stichos*, order.

SYNOPSIS OF THE SPECIES.

§ **Hypopeltis.**—*Pinnæ and pinnules continuous with the rachis.*

* *Fronds pinnate.*

1. **P. Lonchitis**: pinnæ rigid spinosely ciliato-serrate, falcate, auricled; the upper ones usually overlapping.

** *Fronds bipinnate.*

† *Fronds rigid; pinnules sessile, attached by the acute-angled wedge-shaped base, spiny-serrate.*

2. **P. aculeatum**: fronds broad lanceolate; pinnules ovate acute, sub-falcate, auricled; sori infra-medial.

var. **lobatum**: fronds narrow lanceolate; pinnules nearly all confluent, the lowest only auricled.

var. **argutum**: fronds lanceolate; pinnules distinct, long, narrow, *i. e.* linear acute, auricled.

var. **cristatum**: fronds lanceolate; apical lobes of the pinnæ confluent, forming a dilated somewhat crispy termination.

†† *Fronds lax; pinnules (basal ones of lower pinnæ) with obtuse-angled base, attached by a slender pedicel, bristly serrate.*

3. **P. angulare**: fronds broad lanceolate; pinnules short oblong-ovate, subfalcate, auricled; sori terminal or subterminal.

(*a*) *Fronds normal.*

var. **imbricatum**: fronds narrow lanceolate; pinnules roundish-oblong, imbricated, bristle-tipped; rachis proliferous.

var. **rotundatum**: fronds narrow lanceolate; pinnules few, rotundate, flat, obscurely crenate, or sub-entire.

var. **alatum**: fronds lanceolate; pinnules decurrent, the pedicel obliterated by a wing to the rachis.

var. **confluens**: fronds bipinnate below; pinnules (where perfect) linear, acute, auricled, often depauperated or cuneate, the upper confluent into a linear lobato-serrate apex; upper pinnæ linear-falcate, auricled, serrated.

var. **gracile**: fronds bipinnate, ovate-caudate, lax; pinnules narrow oblong or linear, acute, distinct, scarcely auricled, inciso-serrate.

var. **grandidens**: fronds bipinnate, narrow lanceolate; pinnæ irregularly abbreviated, their apex often flabellate; pinnules variable, coarsely inciso-dentate or laciniate, the teeth subulate.

var. **plumosum**: fronds ovate-lanceolate, thin, chartaceous, bipinnate; pinnules deeply inciso-lobate, all acutely aristate-serrate.

var. **proliferum**: fronds ovate-lanceolate, bi-tri-pinnate; pinnules attenuated, distinct, distantly and attenuately lobed; rachis proliferous.

(*b*) *Fronds monstrous or abnormal.*

var. **cristatum**: fronds lanceolate, the apices of the fronds and pinnæ multifid-crisped.

var. **polydactylum**: fronds narrow lanceolate, the apices of the fronds and pinnæ ramose or multifid, plane.

var. **Kitsoniæ**: fronds ramose above, the branches corymbosely tufted; pinnæ dilated-crisped at their apices; pinnules setaceo-serrate.

THE ALPINE SHIELD FERN, or HOLLY FERN.

POLYSTICHUM LONCHITIS.

P. fronds pinnate, narrow linear-lanceolate, rigid; pinnæ falcately lanceolate, acute, spinosely ciliate-serrate, auricled at the base on the upper side, obliquely wedge-shaped or rounded on the lower, the lowest ones often having both an anterior and posterior auricle. [Plate XV.]

POLYSTICHUM LONCHITIS, *Roth, Fl. Germ.* iii. 71. *Deakin, Florigr. Brit.* iv. 89, fig. 1602. *Babington, Man. Brit. Bot.* 4 ed. 423. *Sowerby, Ferns of Gt. Brit.* 30, t. 15. *Newman, Hist. Brit. Ferns,* 3 ed. 103. *Moore, Handb. Brit. Ferns.* 3 ed. 78; *Id., Ferns of Gt. Brit., Nature Printed,* t. 9. *Schott, Gen. Fil.* (t. 9.) *Presl, Tent. Pter.* 82, t. 2, fig. 7. *Fée, Gen. Fil.* 278.

POLYPODIUM LONCHITIS, *Linnæus, Sp. Plant.* 1548. *Bolton, Fil. Brit.* 34, t. 19. *Smith, Eng. Bot.* xii. t. 797. *Flora Danica,* t. 497. *Sturm, Deutschl. Fl. (Farrn.)* i. t. 4.

ASPIDIUM LONCHITIS, *Swartz, Schrad. Journ. Bot.* 1800, ii. 30; *Id., Syn. Fil.* 43. *Smith, Fl. Brit.* 1118; *Id., Eng. Fl.* 2 ed. iv. 271. *Hooker & Arnott, Brit. Fl.* 7 ed. 583. *Mackay, Fl. Hib.* 338. *Bentham, Handb. Brit. Fl.* 628. *Schkuhr, Krypt. Gew.* 29, t. 29. *Willdenow, Sp. Pl.* v. 224. *Sprengel, Syst. Veg.* iv. 97. *Ledebour, Fl. Ross.* iv. 512. *Koch, Synops.* 2 ed. 976. *A. Gray, Bot. North. U. States,* 2 ed. 599. *Fries, Sum. Veg.* 82. *Nyman, Syll. Fl. Europ.* 431. *Lowe, Nat. Hist. Ferns,* vi. t. 22.

ASPIDIUM ASPERUM, *Gray, Nat. Arr. Brit. Pl.* ii. 6.

Caudex thick, slowly elongating, erect or decumbent, consisting of the densely packed bases of decayed fronds surrounding a central woody axis, and clothed in the upper part with the numerous scales which remain about the bases of the stipes. *Fibres* stout, rigid, branched, dark brown.

Vernation circinate.

Stipes usually short, from half an inch to two inches, or sometimes three inches in length, rarely more, clothed with large ovate or broadly-lanceolate reddish-brown pointed chaffy scales; terminal and adherent to the caudex. *Rachis* densely scaly, with narrower lanceolate and subulate pallid scales.

Fronds from six to eighteen, rarely twenty-four inches in length, deep green, paler beneath, of rigid leathery texture, erect or pendulous according to the conditions of growth, linear-lanceolate, pinnate. *Pinnæ* undivided, numerous, with one of the margins, usually the anterior one, bent back from the plane of the rachis, and usually crowded, so that when the frond is flattened they become overlapping on the upper part of the frond, though distinct and sometimes distant below. They are very rigid, with numerous small hair-like scales scattered over their under surface; very shortly stalked or sessile, lanceolate-falcate, from three-quarters of an inch to an inch and a quarter in length in the widest part, having an acute point, and an acute auricle at the base on the anterior side, the base on the posterior side being obliquely sloped or rounded off in all the upper pinnæ, but often produced into a posterior auricle in the lowest ones, which are shorter, and nearly triangular in outline, sometimes even hastate. The margin is serrated, and the serratures are tipped by bristle-like points, with minute intermediate teeth.

Venation generally indistinct. There is a pinnately branched *costa* extending to the apex of the pinna, and diverging from it, at the very point where it enters the pinna, is a principal branch or *vein* which extends to the apex of the auricle, which vein is pinnately-branched on the same plan as the midvein, but on a smaller scale. The rest of the *veins* on each side the midvein are pinnately forked, *i. e.*, they are branched, but the branches are so placed that at each ramification the vein seems to have separated into two nearly equal and but slightly diverging parts. In average specimens there are three or four of these ramifications to each of the veins near the base of the pinna, then two, and finally one in those near the apex. The *venules* and *veinlets* are lost in the substance of the frond just within the margin, one being directed into each marginal tooth. In smaller specimens the number of ramifications in the veins is fewer.

Fructification on the back of the frond, and usually confined to the upper half, though sometimes extending lower down. *Sori* round, indusiate, forming a line on each side the midvein, halfway between it and the margin, and also in a similar way a line on each side the principal vein extending into the auricle. These sori are of variable size, but often large and crowded, and then generally becoming confluent in age; they are attached to the anterior branch of each fascicle of veins, and are medial, seated nearer to its base than its apex. *Indusium,* or cover to the spore-cases, membranaceous, orbicular, umbilicate or peltate, *i. e.* attached to the receptacle by a short central stalk. *Spore-cases* numerous, globose, stalked, deep brown. *Spores* small, round or oblong, granulate.

Duration. The caudex is perennial, and the plant evergreen; the fronds, which appear, as is usual, in the spring months, attain their maturity by the autumn, and remain in full vigour through the winter onwards.

This plant may be taken as the type of *Polystichum,* a genus established by Roth almost contemporaneously with the publication of *Aspidium,* with which in its original form it is synonymous, and which has generally been allowed to supersede it where the genus has been preserved in its entirety. The present species is also the type of *Polystichum* in the restricted sense proposed by Schott, whose views we adopt.

Polystichum Lonchitis is known from perfectly developed states of the cognate species by its being simply pinnate, but from young and imperfect or debilitated forms of the latter, which sometimes occur, and are only pinnate, it is not so readily distinguishable. The rigidity of texture, the strongly spinous margin, and the tendency to imbrication in the pinnæ, offer the readiest marks of distinction from these anomalous congeners which are imperfect forms of *P. aculeatum.*

This species seems less liable to sport into abnormal forms than the others of the genus. It does occur sometimes with the fronds divided at the apex, but this is merely an occasional and accidental variation. The plants, moreover, sometimes produce small bulbils in the axils of the lowermost pinnæ, from which young plants spring up. This quality of producing bulbils seems to be the result in great measure of certain little understood peculiarities of cultivation or situation; for while with some cultivators many of the British species prove bulb-bearing, the peculiarity seldom occurs with others.

This Fern may be considered as an alpine rock-plant. It is plentiful on the mountains of the Scottish Highlands, where it has a range of from about 1200 feet to upwards of 3000 feet (1100 yards, *Watson*) above the sea. It descends to about 1500 feet above the sea in Yorkshire. Its southern ascertained limit occurs in North Wales, in the county of Carnarvon; and it is found again in Yorkshire and the Lake district, and more abundantly in the Highlands of Scotland, its northern reported limit being Ross-shire, Sutherlandshire and Orkney. In Ireland, though the habitats are not numerous, they are widely dispersed. There are dubious reports of its having been met with in the Scottish Lowlands (Lanarkshire), in South Wales (Glamorganshire), and in the district of the Ouse in England (Cambridgeshire, Northamptonshire), but these localities all need confirmation. The recorded stations are:—

S. Wales.—? Glamorganshire.

N. Wales.—Carnarvonshire: Snowdon, Clogwyn-y-Garnedd; above Llanberis; Cwm-Idwal; Twll-du; Moel Hebog; Glyder-Vawr.

Humber.—Yorkshire: Langcliffe, and about Settle; Attermine Scar; Giggleswick; Ingleborough.

Tyne.—Durham: Falcon Clints, Teesdale; Mazebeck Scar.

Lakes.—Cumberland: Fairfield, Helvellyn, *Rev. W. H. Hawker.*

W. Lowlands.—? Lanarkshire.

E. Highlands.—Stirlingshire: Ben Lomond, *F. Bossey.* Perthshire: Ben Lawers; Craig Challiach; Glen Lyon, *G. Lawson;* Ben Chonzie, near Crieff, *Dr. Balfour;* Ben Voirlich; Ben Ledi, and Callender, *Mrs. Macleod.* Forfarshire: Canlochen, Glen Isla, Glen Fiadh, Craig Maid, Glen Dole, etc., in the Clova Mountains, Aberdeenshire. Morayshire.

W. Highlands.—Inverness-shire: Mountains near Loch Ericht. Isle of Mull: Ben More.

N. Highlands.—Ross-shire: Raven Rock, near Castle Leod. Sutherlandshire: Ben Hope; Assynt.

N. Isles.—Orkney: Hoy-hill (1600 feet), very rare, *T. Anderson.*

Ulster.—Donegal: Glen E. of Lough Eske; Rosses and Thanet Mountain passes, *D. Moore.*

Connaught.—Sligo: Ben Bulben. Leitrim: Glenade Mountains.

Leinster.—Meath: Navan, *R. Kyle.*

Munster.—Kerry: Brandon Hill; Mangerton.

The Holly Fern would appear to be extensively distributed over the great mountain regions of Europe, especially in the northern and central portions, occurring in Iceland and Lapland; in Sweden, Denmark, and Arctic Russia; in Great Britain and Ireland; in Germany, Hungary, Dalmatia, Croatia, and Transylvania; in France, Belgium, and Switzerland, extending to Italy and Spain, and southwards to Crete, and Mount Taygetus in Greece. In Asia, it occurs in Kashmir, as well as on the Bithynian Olympus, and on the Russian Altai range, extending into the Arctic regions as far as Kamtchatka; thence it passes to the Rocky Mountains in North-West America, and to Disco Island in Davis's Strait. A closely-related species, the *Aspidium munitum* of Kaulfuss, intermediate between our *P. Lonchitis* and the *P. falcinellum* of Madeira, is found in California, and at Nootka Sound.

Our own experience of the cultivation of *Polystichum Lonchitis* is, that it is a plant of shy growth, and very tardy increase. It requires a cool moist shady frame, and when once established may be kept in good condition in such a situation. It must be potted, very firmly, in well drained loamy gritty soil, and be kept freely supplied with moisture, which however on no account should be stagnant. It seems to need a pure air, as its mountain home would suggest, and probably does not like the denser atmosphere of low-land situations. Hence in confined or smoky localities it will not long exist on out-door rockwork, but in places where the atmosphere is salubrious, and especially if the situation is elevated, it will succeed in a shady rockery, where the damp but well drained conditions of its natural localities, can be tolerably closely imitated. The climate of Ireland seems more congenial to this plant, than that of the south of England.

The fact that this species, though a vigorous looking and hardy plant in its native haunts, is seldom seen to preserve its vigour under cultivation, at least in the neighbourhood of London, is mainly, we think, owing to the impossibility of securing the pure atmosphere of its native mountains. It certainly prefers a damp atmosphere; and, provided the moisture is not stagnant, its roots too should be freely supplied; they should in fact be constantly damp with percolating moisture. Hence the necessity of a careful mechanical adjustment of the materials employed as compost, of which mellow loam, gritty sand, and small masses of some porous body, such as soft sandstone, should be the main ingredients. We have succeeded tolerably well by potting the plants very firmly in a compost formed of materials such as those just indicated, with a small proportion of peat added, the plants being kept under glass in a close-shaded cold frame; but found on removal for the winter to a cold greenhouse where the atmosphere was drier and less confined, that the fronds were considerably injured by the exposure. The same plants however on being enclosed within a handglass, where consequently the atmospheric moisture was more abundant and regular, grew vigorously. We have therefore no doubt that the requirements of this species are, a well moistened but freely-drained soil, and a damp atmosphere—pure if possible; and these conditions can only be secured, in many cases, by keeping the plants close under glass in a north aspect. Propagation is rarely to be effected by division, lateral crowns being seldom produced; consequently, plants have generally to be obtained from their native habitats.

It is of some importance to bear in mind, in attempting to remove shy-growing species like the present from their natural habitats to the garden, that the older and larger plants are generally less successfully removed than the younger and smaller ones. One of the first conditions of success undoubtedly is to remove them with the least possible injury to their roots, and this is more likely to be effected in the case of small plants, than with those which are more thoroughly established, especially if the situation in which they are growing is rocky.

THE COMMON PRICKLY SHIELD FERN.

POLYSTICHUM ACULEATUM.

P. fronds bipinnate, lanceolate or broad linear-lanceolate, rigid; pinnules ovate-subfalcate, auricled, acute, distinct and attached by their wedge-shaped base, or obliquely decurrent, or confluent, the anterior basal ones larger, all prickly-serrate; sori infra-medial. [Plate XVI.]

POLYSTICHUM ACULEATUM, *Roth, Fl. Germ.* iii. 79. *Deakin, Florigr. Brit.* iv. 91, fig. 1603. *Babington, Man. Brit. Bot.* 4 ed. 423. *Sowerby, Ferns of Gt. Brit.* 32, t. 17 (incorrect as to venation). *Newman, Hist. Brit. Ferns,* 3 ed. 169, in part. *Moore, Handb. Brit. Ferns,* 3 ed. 81; *Id., Ferns of Gt. Brit. Nature Printed,* t. 10. *Schott, Gen. Fil.* (t. 9.) *Presl, Tent. Pterid.* 83. *Fée, Gen. Fil.* 278.

POLYSTICHUM LOBATUM, *Presl, Tent. Pterid.* 83. *Link, Fil. Sp.* 111. (excl. var. syn.) *Hooker, Gen. Fil.* t. 48 C. *Fée, Gen. Fil.* 278.

POLYSTICHUM AFFINE, *Presl, Tent. Pterid.* 83.

POLYPODIUM ACULEATUM, *Linnæus, Sp. Plant.* 1552.

ASPIDIUM ACULEATUM, *Swartz, Schrad. Journ. Bot.* 1800, ii. 37; *Id., Syn. Fil.* 53. *Smith, Fl. Brit.* 1122; *Id., Eng. Bot.* xxii. t. 1562; *Id., Eng. Fl.* 2 ed. iv. 277. *Hooker & Arnott, Brit. Fl.* 7 ed. 583. *Bentham, Handb. Brit. Fl.* 628. *Willdenow, Sp. Plant.* v. 258. *Sprengel, Syst. Veg.* iv. 105. *Ledebour, Fl. Ross.* iv. 512 (excl. syn. *vestitum* and *setigerum*). *Koch, Synops.* 2 ed. 976. *Sturm, Deutschl. Fl.* (*Farrn.*) t. 3. *Tenore, Att. Accad. R. Inst. Sc. Nat. Nap.* v. (reprint 22, t. 2, fig. 5). *Spenner, Fl. Friburg.* i. 9, t. 1. *Nyman, Syll. Fl. Europ.* 431.

ASPIDIUM LOBATUM, *Schkuhr, Krypt. Gew.* 42, t. 40. *Kunze, Flora,* 1848, 356.

ASPIDIUM DISCRETUM, *Don, Prod. Fl. Nep.* 4.

ASPIDIUM AFFINE, *Wallich, Cat.* 370.

Var. **lobatum**: fronds narrow-lanceolate, very rigid; pinnules (the larger basal ones only, distinct and auricled) elliptic, convex, not auricled, nearly all decurrent or confluent, prickly-serrate. [Plate XVII A.]

POLYSTICHUM ACULEATUM *v.* LOBATUM, *Deakin, Florigr. Brit.* iv. 91. *Moore, Handb. Brit. Ferns,* 2 ed. 86; 3 ed. 81; *Id., Ferns of Gt. Brit. Nature Printed,* t. 11. *Babington, Man. Brit. Bot.* 4 ed. 423. *Fée, Gen. Fil.* 278.

POLYSTICHUM ACULEATUM *v.* LONCHITIDOIDES, *Deakin, Florigr. Brit.* iv. 91.

POLYSTICHUM ACULEATUM, *Link, Fil. Sp.* 111. (excl. var. syn.). *Newman, Hist. Brit. Ferns,* 3 ed, 169, in part.

POLYSTICHUM LOBATUM, *J. Smith, Hook. Journ. Bot.* iv. 195. *Sowerby, Ferns of Gt. Brit.* 33 t. 16.

POLYSTICHUM PLUKENETII, *De Candolle, Fl. Franç.* v. 241.

ASPIDIUM LOBATUM, *Swartz, Schrad. Journ. Bot.* 1800, ii. 37; *Id., Syn. Fil.* 53. *Smith, Fl. Brit.* 1123; *Id., Eng. Bot.* xxii. t. 1563; *Id., Eng. Fl.* 2 ed. iv. 278. *Hooker & Arnott, Brit. Fl.* 7 ed. 583. *Mackay, Fl. Hib.* 338. *Willdenow, Sp. Plant.* v. 260. *Tenore, Att. Accad. R. Inst. Sc. Nat. Nap.* v. (reprint 24, t. 2, fig. 6).

ASPIDIUM LOBATUM, *β.* LONCHITIDOIDES, *Hooker & Arnott, Brit. Fl.* 7 ed. 583.

ASPIDIUM ACULEATUM, *Schkuhr, Krypt. Gew.* 41, t. 39. *Lowe, Nat. Hist. Ferns,* vi. t. 16.

ASPIDIUM PLUKENETII, *Steudel, Nom. Bot.* ii. 64.

ASPIDIUM INTERMEDIUM, *Sadler, Adumb. Epiphyll. Hung.* 16.

ASPIDIUM MUNITUM, *Sadler, Fil. Hung.* 54; not of Kaulfuss.

POLYPODIUM LOBATUM, *Hudson, Fl. Ang.* 459.

POLYPODIUM ACULEATUM, *Bolton, Fil. Brit.* 48, t. 26.

POLYPODIUM PLUKENETII, *Loiseleur, Not.* 146.

Var. **argutum**: fronds lanceolate; pinnules distinct, long, narrow, *i. e.* linear acute, auricled, sharply spine-toothed. [Plate XVII B.]

POLYSTICHUM ACULEATUM, *v.* ARGUTUM, *Moore, Ferns of Gt. Brit. Nature Printed,* t. 10 B; *Id., Handb. Brit. Ferns,* 3 ed. 82.

Var. **cristatum**: fronds lanceolate, tapered to the point; pinnæ somewhat dilated and crispy at their apices, the upper ones confluent; pinnules oblong acute, auricled.

Caudex thick, tufted, erect or decumbent, becoming woody in age, consisting of the bases of decayed fronds closely surrounding a woody axis, slowly elongating, in the upper part scaly. *Scales* broad ovate-lanceolate, numerous, dark fuscous. *Fibres* long, coarse, tortuous, branched, dark brown, tomentose.

Vernation circinate, the main rachis becoming recurved before the unfolding of the frond is completed; the pinnæ convolute towards the main rachis.

Stipes short, three to four inches long, densely scaly with broad ovate-lanceolate chaffy fuscous scales; terminal and adherent to the caudex. *Rachis* stout, rounded behind, channelled in front, densely scaly, the scales less numerous and hair-like above, more numerous and intermixed with broader ones below, gradually merging in size with those of the stipes.

Fronds from one to three feet high, and from four to seven inches across, rigid, leathery, smooth and dark-green above, paler beneath, erectish or more or less spreading, occasionally somewhat drooping, lanceolate in form, bipinnate. *Pinnæ* numerous, obliquely-lanceolate, broadest at the base, acuminate, pinnate at the base and for a part of their length, sometimes nearly to the apex, in other cases the basal pinnules only being distinct; the upper ones alternate, the lower ones nearly opposite and diminishing in size. *Pinnules* ovate-falcate or elliptic, acute and aristate at the apex; all or the basal ones only auriculate on the anterior side, the auricle acute and mucronate; aristate, subsessile, and attached by the wedge-shaped base, or decurrent; the basal portion entire, and when distinct, obliquely incised on the posterior side, truncate on the side next their rachis; the rest of the margin toothed with unequal adpressed mucronate serratures. The basal anterior pinnule on each pinna is generally larger, often much larger than the rest, and more strongly auricled, and the pinnules are all more or less convex; on the under surface are scattered fine hair-like scales. The typical form of the species has the pinnules mostly distinct; the variety *lobatum* has them mostly decurrent, while in some plants of the latter, apparently resulting indifferently from youth or decrepitude, they are obsolete, the pinnæ being merely more or less deeply lobed and toothed, somewhat resembling those of *P. Lonchitis,* and hence specimens in this state are sometimes named *lonchitidoides.*

Venation of the pinnules consisting of a flexuous *costa* or midvein, with alternate branches or *veins,* which are again furcately-branched alternately, the lower veins producing three or four, the upper two or three branches or *venules,* of which the lowest anterior one of the fascicle is soriferous. In the auriculate portion at the base, the vein is more prominent than in the upper portion, and gives off a greater number of simple or forked venules, some few of which on both sides may produce sori.

Fructification on the back, and usually confined to the upper half of the frond. *Sori,* round, indusiate, seated much below the apices of the venules, in a line on each side of the midvein of the pinnules, and also of the vein of the auricles; often crowded, sometimes becoming confluent; attached to the lowest anterior venule of the fascicle of veins, or, at the auriculate base, to the venules on either side the vein; but there also to the anterior branch if they are forked. *Indusium* membranaceous, orbicular, peltately attached. *Spore-cases* numerous, dark-brown, roundish-obovate, stalked. *Spores* slightly muriculate.

Duration. The caudex is perennial. The fronds are persistent through the winter and the following summer, though sometimes damaged by severe frosts. The young fronds grow up in May.

The division of the bipinnate aculeate Ferns into three species, i. e. *lobatum*, *aculeatum*, and *angulare*, Mr. Newman observes,* probably originated in an error of nomenclature, and he arrives at this conclusion from a careful consideration of the original descriptions. "Linnæus considered the plants referrible to single species to which he gave the name *aculeatum*. Hudson, observing the great discrepancy between the extreme forms, divided them into two species, calling the rigid and least divided form *lobatum*, and the lax and most divided form *aculeatum*. Kunze adopted these names; but Willdenow had redescribed the species, transferring the name *aculeatum* to Hudson's *lobatum*, and giving the new name of *angulare* to Hudson's *aculeatum*. Thus the three names were not intended to represent three objects; a conclusion inadvertantly adopted by Sir J. E. Smith. There is now a growing disposition to reunite them as one species." It is however doubtful whether Linnæus knew anything of *angulare*, though there is hardly room to doubt that he included the other two forms under *aculeatum*, which is the view which both ourselves and Mr. Newman have adopted.

Though *P. aculeatum* is often difficult to distinguish from our next species *P. angulare*, yet viewing the British forms alone, it appears to be really distinct. Indeed, if these plants are not distinct, a series varying through every gradation from the pinnate *lonchitidoides* to the tripinnate forms of *angulare* must be united, and all hope of defining a species would then be at an end. While admitting the difficulty of discriminating between some forms, especially exotic ones, of these two species, we may endeavour to point out how they may generally, with tolerable certainty, be known from each other.

* Newman, *History of British Ferns*, 3 ed. 112.

(1) *P. aculeatum* is stouter, more erect, and altogether more rigid in texture; while *P. angulare* is normally lax and more herbaceous, and equally large or even larger in size.

(2) *P. aculeatum* has its pinnules either confluent or decurrent, (in which cases there is no difficulty whatever in distinguishing it), or, when the pinnules are distinct, as in the most perfect plants, they are wedged-shape at the base, the anterior side being truncate, and the posterior obliquely incised in straight lines, the two lines describing an acute angle by the apex of which they are attached to the rachis; while in *P. angulare* the truncated anterior base is more curved in outline, and the two lines of the base describe a right angle or an obtuse angle, at the apex of which is a distinct slender petiole, by which they are attached.

(3) *P. aculeatum* has its sori medial, that is, attached at a point along the middle part of the venule, the apex of which is carried out to the margin of the pinnule, the sori thus being placed nearer the base of the venule than its apex, *i. e.*, nearer the point of furcation; while in *P. angulare* the fertile venule stops about midway across the pinnule, and the sorus is commonly placed at or almost close to its apex. These peculiarities observed in connection with each other will serve to reduce the dubious forms within very narrow limits indeed, at least so far as British examples are concerned. The portion rather below the middle of the frond and the basal pinnules should be taken for examination.

This is one of the most easily cultivated of all the larger hardy Ferns. It prefers a loamy soil and partial shade; and is increased readily by division. Being evergreen, its varieties are among the most desirable of our native species for the decoration of shady walks and rockeries, in which latter situation especially, where the roots are generally well drained, provided the plants are not exposed to the effects of severe drought and are moderately shaded, they thrive admirably. It is also very manageable as a pot plant, and under any circumstances is ornamental in its character. The smaller form, known as *lobatum*, is perhaps the most suitable for pot culture, on account of its size and the elegance of its fronds, which not uncommonly assume a very graceful lateral curve. The plants, though a good deal damaged in severe winters if exposed, are thoroughly evergreen under shelter.

The Common Prickly Shield Fern is a widely-dispersed and not uncommon plant, in shady hedge-banks, woods, and similarly sheltered situations. It is found all over England, Scotland, and Wales, in the majority of the counties, the records of the variety *lobatum* showing an equally wide, if not wider range. The four provinces of Ireland also produce it, as do the Channel Isles; but we are not aware of its having been found in the Northern or Western Isles. It occurs at the sea-level, in the south-western parts of England; and in the form of *lobatum*, which seems the more common in Scotland, it ascends to upwards of 2000 feet in the Highlands. The records of its distribution, so far as we have ascertained, are as follows, but this species and *P. angulare* not being always well distinguished, the particulars are doubtless imperfect, and perhaps in some cases incorrect:—

Peninsula.—Cornwall. Devonshire: Lynmouth; Kingsteignton; between Totnes and Ashburton; Barnstaple, *F. Mules.* Somersetshire: Portishead; Dundry Hill, near Bristol, *G. H. K. Thwaites*; Nettlecombe.

Channel.—Hampshire: Maple-Durham, near Petersfield, *J. Goodyer*, July, 1633; Selborne, *Miss Bower*; Bramshot; Alresford, etc. Isle of Wight: Gurnet Bay. Dorsetshire. Wiltshire: Box quarries, *R. Withers*; Bretford, Salisbury; and Redlynch, Downton, *W. Moore.* Sussex: Henfield; Cuckfield; Hastings, etc.

Thames.—Hertfordshire: St. Alban's; Totteridge; Hitchin; Essendon; Cheshunt, etc. Middlesex: Norwood, *S. F. Gray*; Osterley Park; Lampton lane; Hendon, *W. Pamplin.* Kent: Tunbridge Wells. Surrey: Ockshot, *H. C. Watson*; Mayford; Dorking; Denham, *Mrs. James*, and elsewhere. Buckinghamshire: Fulmer. Berkshire: Swallowfield, *R. Heward.* Oxfordshire. Essex: near Ongar; Brentwood; Chingford; Little Warley Common.

Ouse.—Suffolk: Sudbury, etc. Norfolk: Edgefield, near Holt. Cambridgeshire: Gamlingay. Bedfordshire. Northamptonshire.

Severn.—Warwickshire: Stoneleigh; Allesley; Rugby, *Rev. A. Bloxam;* Hollyberry End, and Wyken-lane; Alcester (with tasselled fronds), *Hb. Hooker;* Elmdon House, and elsewhere. Monmouthshire: Mamhilad, *T. H. Thomas.* Herefordshire. Gloucestershire: near Bristol; Leigh; Stapleton; Shapscombe Wood, near Painswick; Foscott, and Broadwell, *H. Buckley.* Worcestershire: Knightwick, *E. Lees;* Bromsgrove. Staffordshire. Shropshire: Bridgenorth; Mannington, near Cherbury.

Trent.—Leicestershire: Charnwood Forest. Nottinghamshire: Beeston; Mansfield; Paplewick. Derbyshire: Matlock. Lincolnshire.

Mersey.—Lancashire: Chaigeley, near Clitheroe, *E. J. Lowe;* Manchester; Gateacre, near Liverpool; Hail Wood; Burton Wood, near Warrington, etc. Cheshire: Preston.

Humber.—Yorkshire: Sowerby Dene, near Halifax; Castle Howard Woods; Settle; Richmond; Studley; Roche Abbey, *J. F. Young;* Ripon; Bradford, *J. T. Newboult;* Doncaster; Sheffield; York; Ingleborough.

Tyne.—Northumberland: Hexham, etc. Durham: Cawsey Dene, etc., *R. Bowman.*

Lakes.—Cumberland: Irton Wood, *J. Robson;* Airey Force, *H. Fordham,* etc. Westmoreland: Ambleside; Rydal. Isle of Man.

S. Wales.—Pembrokeshire: Tenby, *E. Lees;* Castle Malgwyn, Llechryd, *W. Hutchison.* Carmarthenshire. Glamorganshire. Brecknockshire: Talgarth, *E. Williams,* and elsewhere common, *J. R. Cobb.*

N. Wales.—Anglesea: Lleiniog Castle; Cickle, near Beaumaris. Denbighshire: Wrexham; Ruthin; Llanymyneck. Carnarvonshire: Llyn-y-cwm; Bangor.

W. Lowlands.—Dumfriesshire: Drumlanrig; Nithsdale, and other parts, *P. Gray.* Kirkcudbrightshire, *P. Gray.* Renfrewshire. Lanarkshire: Cartland Rocks.

E. Lowlands.—Edinburghshire: Hawthornden, *T. M.* Berwickshire: Pease Bridge, etc.

E. Highlands.—Forfarshire. Fifeshire: St. David's. Perthshire: Dunkeld, *A. Tait;* Glenfarg, near Perth. Kincardineshire. Aberdeenshire.

W. Highlands.—Argyleshire: Glen Gilp, Ardrishiag. Isles of Islay, Cantyre, and Bute.

N. Highlands.—Ross-shire.

Ulster.—Antrim: Colin Glen, *Dr. Mateer;* Malone, near Belfast; Carrickfergus.

Connaught.—Galway: Connemara; Gort, *J. R. Kinahan.*

Leinster.—Wicklow: Newtown Mount Kennedy. Dublin: Bohernabreena, *J. R. Kinahan.*

Munster.—Clare. Cork: Clonmel; Glendine, near Youghal.

Channel Isles.—Jersey.

This Fern appears to be found over nearly the whole of Europe, its occurrence being recorded, northwards in the Scandinavian kingdoms; through Central Europe, *e. g.* Switzerland, Germany, Holland, Belgium, and France, to the Spanish Peninsula; and extending from Italy, to Dalmatia, Croatia, Transylvania, Greece and Turkey. In Asia, it is found in the Russian dominions, from Colchis to Lenkoran; and in various parts of British India—the *Aspidium discretum* of Don being the same with the British *aculeatum*, as appears from authentic specimens in the Herbarium of the Linnæan Society. In Africa it occurs on the northern coast, at Algiers, and again in the south, as well as in the Island of Madeira. In America, its range extends from the Eastern United States to Columbia on the north-west coast. The Indian *Aspidium lentum* of Don (*A. ocellatum*, Wall.) has a very close resemblance to the variety *lobatum*, but being freely proliferous, is perhaps distinct. There exist, moreover, several South American Ferns, which, if not specifically identical with the European *P. aculeatum*, are at least very intimately allied to it; such specimens have been gathered in Mexico, in Guatemala, and in Columbia. Others from Brazil, the *Polystichum microphyllum* of Klotsch, seem to be referrible here; as also does the *Aspidium subintegerrimum* of Hooker and Arnott, from Chili. In the same doubtful category must be placed the *Aspidium vestitum;* the *Aspidium venustum* of Hombron and Jacquinot, from the Island of Auckland, and New Zealand; the *Aspidium proliferum* of Tasmania; and the *Aspidium setosum*, and *Aspidium rufo-barbatum* which occur in various parts of the East Indies.

There are various degrees of development in this species, some of the most distinct of which have been considered as varieties. One at least of them (*lobatum*) has been distinguished as a species by various authors; and another, *lonchitidoides*, has been sometimes regarded as a distinct variety, sometimes as the young state of *lobatum*. We are of opinion, that *lobatum* may be considered as a variety of *aculeatum* without violence to nature; and that *lonchitidoides*, rather than a distinct variety, is the partially developed or debilitated condition of *lobatum*, whether caused by youth, age or starvation, or any other depressing influence. The two are certainly not permanently distinct, but interchangeable, for cultivated plants of *lonchitidoides* may be nurtured onwards into *lobatum* proper, and *lobatum* may be starved back into *lonchitidoides*. The plant to which the latter name is given, is a dwarf, simply pinnate, fertile form, often very much resembling *P. Lonchitis*, but less spiny, not imbricated, and with a greater or less tendency to become lobed. The species has occasionally been found with the apex multifid and the pinnæ dichotomous; and sometimes has been known to produce bulbils in the axils of the lower pinnæ; but these variations are accidental. The more marked varieties are:—

1. *lobatum* (Deak.). This variety is doubtless a more fully developed condition of the lonchitidoid form above mentioned. It has narrow lance-shaped fronds, one to two feet long; these are subbipinnate, *i.e.*, a few only of the pinnæ develope pinnules. The anterior basal pinnule is always distinct, considerably enlarged, and strongly auricled; but the rest of the pinnules are either decurrent or confluent, and not auricled. The type form of the species is broader, and most of its pinnules are distinct and auricled, and between this and the variety *lobatum*, there is to be found numerous intermediate grades; but yet our experience does not tend to the conclusion that the marked forms of *lobatum* can be made to develope into *aculeatum* by culture, but on the contrary, that it is a permanent variety of which various gradations exist in a natural state. [Plate XVII A.—Folio ed. t. XI.]

This plant (var. *lobatum*) is by no means uncommon, as will appear from the following recorded habitats:—

Peninsula.—Devonshire: Totnes; Barnstaple, *F. Mules*; Challacombe, *F. Mules* (lonchitidoid form); Tiverton, *Miss Hutton* (lonchitidoid form). Somerset: Portishead; Nettlecombe.

Channel.—Hampshire: Selborne. Isle of Wight. Wiltshire: Box quarries, *T. Z. Lawrence* (lonchitidoid form). Sussex: Cuckfield; Groombridge.

Thames.—Middlesex: Norwood, *S. F. Gray.* Kent: St. Mary Cray, *R. Sim.* Surrey: Mayford, *T. M.*; Dorking, etc. Buckinghamshire: Chalfont. Berkshire. Oxfordshire. Essex: Black Notley; Norton Heath; Chingford.

Ouse.—Suffolk: Winkfield; Spexhall; Sudbury. Norfolk: Yarmouth; Tivetshall, *Miss Wells.* Northamptonshire.

Severn.—Warwickshire: Stoneleigh; Allesley, *Rev. W. Bree*; Rugby; Wyken-lane; Studley; Overley; Weatherley, *Rev. W. Bree.* Monmouthshire: Mamhilad. Herefordshire (lonchitidoid form). Gloucestershire: near Bristol; Broadwell, *H. Buckley*; Littleworth. Staffordshire. Shropshire: Bridgenorth (lonchitidoid form); Mannington, near Cherbury, *Rev. W. M. Hind* (lonchitidoid form); Blodwell Rocks, *Rev. W. A. Leighton* (lonchitidoid form).

Trent.—Leicestershire. Nottinghamshire: Paplewick. Derbyshire: Matlock. Lincolnshire.

Mersey.—Lancashire: Chaigeley near Clitheroe; Walton; Manchester; Hail Wood. Cheshire: Preston.

Humber.—Yorkshire: Fountain's Abbey; Ripon, *A. Clapham* (small divided form); Heckfell Woods; Sheffield; Ingleborough; York; Richmond; Studley; Scarborough, *A. Clapham* (lonchitidoid form); Halifax; Settle; Pottery Car, near Doncaster; etc.

Tyne.—Northumberland: Hexham; Scotswood Dene (furcate). Durham: Cawsey Dene, etc.

Lakes.—Cumberland: Irton Wood; Airey Force. Westmoreland: Patterdale (lonchitidoid form).

S. Wales.—Pembrokeshire. Glamorganshire. Brecknockshire: Talgarth; Llandrindod Wells, *Rev. T. Salwey* (lonchitidoid form).

N. Wales.—Anglesea. Denbighshire: Ruthin; Rhuabon, *A. L. Taylor* (lonchitidoid form); Llanymyneck; Wrexham. Carnarvonshire: Llanberis.

W. Lowlands.—Dumfriesshire: Nithsdale, etc. Kirkcudbrightshire. Lanarkshire.

E. Lowlands.—Edinburghshire: Braid Woods. Berwickshire: Pease Bridge.

E. Highlands.—Forfarshire: Glen Fiadh, Clova, *W. Wilson* (lonchitidoid form). Perthshire: Trosachs, *T. M.;* Dunkeld. Kincardineshire. Aberdeenshire. Nairnshire: Cawdor Woods. Morayshire.

W. Highlands.—Argyleshire: Glen Gilp, Ardrishiag, *T. M.* Isles of Islay, and Cantyre.

N. Highlands.—Ross-shire.

Ulster.—Antrim: Colin Glen; Malone, near Belfast (lonchitidoid form). Londonderry, *D. Moore.*

Leinster.—Wicklow: Newtown Mount Kennedy, *R. Barrington;* Hermitage Glen, *Dr. Osborne.*

Munster.—Clare: foot of 'Mononita' (lonchitidoid form).

Connaught.—Sligo, *Mrs. Barrington* (lonchitidoid form).

2. *argutum* (M.). This variety has a broad lanceolate frond, with distinct pinnules as in the typal plant, from which it differs in the form of the pinnules, which are narrowed and elongated, becoming linear terminating in an acute spiny point, and having long spines to the marginal teeth and a prominent auricle. It was gathered in some part of Buckinghamshire, by Mr. J. Lloyd. [Plate XVII B.—Folio ed. t. X B.]

3. *cristatum* (M.). This variety, which has only recently been discovered, has the pinnæ pinnate for about half their length, while the apical portion is pinnatifid with oblong acute lobes, rounded in front as in *lobatum,* these lobes becoming more and more confluent as they approach the apex which is slightly dilated and crispy as in the cristate varieties of other Ferns, but in a less degree. The upper pinnæ are confluent, and the apex of the frond is acuminate. The pinnules are oblong, acute, with a wedge-shaped base and a small anterior auricle. It was found at Barnstaple, by Mr. H. F. Dempster.

4. *crassum* (M.). This variety has the pinnules remarkably thick; they are also short, broad, overlapping, and doubly serrate. It was found near Basingstoke, by Mr. F. Y. Brocas.

5. *multifidum* (Woll.). This has the apex of the frond divided into a spreading tuft of branches. Mr. Wollaston has obtained it from Suffolk. Some plants of this character are inconstant.

THE SOFT PRICKLY SHIELD FERN.

POLYSTICHUM ANGULARE.

P. fronds lanceolate, lax, herbaceo-chartaceous, bipinnate; pinnules distinct, oblong or ovate-subfalcate, auricled, bluntish or acute, with an obtuse-angled base, attached by a distinct stalk, lobed or serrated, the serratures tipped by soft bristles; sori terminal or subterminal. [Plate XVIII.]

POLYSTICHUM ANGULARE, *Presl, Tent. Pterid.* 83. *Newman, Hist. Brit. Ferns,* 3 ed. 117. *Deakin, Florigr. Brit.* iv. 95, fig. 1604. *Babington, Man. Brit. Bot.* 4 ed. 423. *Moore, Handb. Brit. Ferns,* 3 ed. 85; *Id., Ferns of Gt. Brit. Nature Printed,* t. 12 A. *Sowerby, Ferns of Gt. Brit.* 34, t. 18. *Fée, Gen. Fil.* 278.

POLYSTICHUM ACULEATUM, *A. Gray, Man. Bot. North. U. States,* 632. *Fée, Gen. Fil.* 278.

POLYSTICHUM SETIFERUM, *Moore, Ferns of Gt. Brit. Nature Printed,* t. 12, in obs.

POLYSTICHUM AFFINE, *Wollaston, Phytol.* n. s., i. 439.

POLYSTICHUM ORBICULATUM, *Gay, Hist. Chil.* vi. 515.

POLYSTICHUM BRAUNII, *Fée, Gen. Fil.* 278. (*A variety.*)

ASPIDIUM ANGULARE, *Kitaibel MS.*: *Willdenow, Sp. Plant.* v. 257. *Smith, Eng. Fl.* 2 ed. iv. 278. *Sowerby, Supp. Eng. Bot.* t. 2776. *Hooker & Arnott, Brit. Fl.* 7 ed. 584. *Mackay, Fl. Hib.* 339. *Fries, Sum. Veg.* 82, 252. *Lowe, Nat. Hist. Ferns,* vi. tt. 23, 24. *Nyman, Syll. Fl. Europ.* 431.

ASPIDIUM ACULEATUM, *Kunze, Flora,* 1848, 359 (excl. var. syn.); *Id., Lin.* xxiii. 224.

ASPIDIUM ACULEATUM, β., *Smith, Fl. Brit.* 1122. *Bentham, Handb. Brit. Fl.* 629. *Ledebour, Fl. Ross.* iv. 513.

ASPIDIUM ACULEATUM, γ. BRAUNII, *Döll, Rhein. Fl.* 21. *A. Gray, Bot. North. U. States,* 2 ed. 599. (*A variety.*)

ASPIDIUM HASTULATUM, *Tenore, Sem. H. R. Neap.* 1830; *Id., Fl. Napol.* iv. 139; v. 304, t. 250, fig. 1; *Id., Att. Accad. R. Inst. Sc. Nat. Nap.* v. (reprint 26, t. 4, fig. 7).

ASPIDIUM LOBATUM, *Loudon, Encyc. of Plants,* 884.

ASPIDIUM LOBATUM, *v.* ANGULARE, *Mettenius, Fil. Hort. Bot. Lips.* 88.

ASPIDIUM ORBICULATUM, *Desvaux, Berl. Mag.* v. 321.

ASPIDIUM FUSCATUM, *Willdenow, Sp. Plant.* 256 (excl. syn.)

ASPIDIUM PAUCICUSPIS, *Sturm, Enum. Fil. Chil.* 33.

ASPIDIUM BRAUNII, *Spenner, Fl. Friburg.* i. 9, t. 2. *Kunze, Flora,* 1848, 362. (*A variety.*)

POLYPODIUM SETIFERUM, *Forskal, Fl. Ægypt. Arab.* 185.

POLYPODIUM APPENDICULATUM, *Hoffmann, Deutschl. Fl.* ii. 8, (not Swartz).

POLYPODIUM ANGULARE, *Fries, Novit. Fl. Succ. Mant.* i. 20; according to Kunze.

POLYPODIUM ACULEATUM, *Hudson, Fl. Ang.* 459. *Bory, Ess. Isles Fort.* 311.

HYPOPELTIS LOBULATA, *Bory, Exp. de Morée,* 286; according to Kunze.

TECTARIA ELONGATA, *Cavanilles, Ann. Cienc. Nat.* iv. 101; according to Webb.

Var. **imbricatum**: fronds linear-lanceolate; pinnæ short, bluntish; pinnules roundish or oblong, obtuse, bristle-tipped, imbricated; rachis proliferous below. [Plate XXI.]

POLYSTICHUM ANGULARE, *v.* IMBRICATUM, *Moore, Ferns of Gt. Brit. Nature Printed,* t. 12 E; *Id., Handb. Brit. Ferns,* 3 ed. 85.

Var. **rotundatum**: fronds narrow lanceolate; pinnæ short, blunt; pinnules few, round or roundish-oblong, obscurely crenate or subentire, flat.

Var. **alatum**: fronds lanceolate; pinnules decurrent with the winged secondary rachides, their teeth rounded, and bristle-pointed. [Plate XXII.]

POLYSTICHUM ANGULARE, *v.* ALATUM, *Moore, Ferns of Gt. Brit. Nature Printed,* t. 10 C, in expl. (*aculeatum* on plate); and under t. 12; *Id., Handb. Brit. Ferns,* 3 ed. 86.
POLYSTICHUM ACULEATUM, *v.* ALATUM, *Moore, Handb. Brit. Ferns,* 2 ed. 86.

Var. **confluens**: fronds bipinnate below; pinnules narrow, the perfect ones linear acute, with an auricle, but the majority more or less depauperated, often cuneate, the upper ones confluent into a linear lobato-serrate apex; upper pinnæ linear-falcate auricled, serrate. [Plate XXVI.]

POLYSTICHUM ANGULARE, *v.* CONFLUENS, *Moore, Handb. Brit. Ferns,* 3 ed. 94.

Var. **gracile**: fronds ovate-caudate, bipinnate, lax; pinnules narrow oblong or linear, acute, scarcely auricled, distinct, inciso-serrate. [Plate XXV B.]

Var. **grandidens**: fronds bipinnate, narrow lanceolate, the pinnæ irregularly abbreviated, often terminating in an abrupt flabellately-lobate apex; pinnules short, very coarsely inciso-dentate, the teeth subulate. [Plate XXV A.]

POLYSTICHUM ANGULARE, *v.* GRANDIDENS, *Moore, Handb. Brit. Ferns,* 3 ed. 94.

Var. **plumosum**: fronds ovate-lanceolate, thin chartaceous, bipinnate; pinnules long stalked, deeply inciso-lobate, the lowest auriculiform lobe distant almost stalked lobed or biserrate; the lobes all acutely aristate-serrate. [Plate XX F.]

POLYSTICHUM ANGULARE, *v.* PLUMOSUM, *Moore, Handb. Brit. Ferns,* 3

Var. **proliferum**: fronds ovate-lanceolate, lax, bi-tri-pinnate; pinnules narrow, attenuated, acute, distinctly stalked, usually deeply-lobed, the lobes distant, attenuated; rachis proliferous throughout. [Plate XXIII.=subvar. *Wollastoni.*]

Polystichum angulare, *v.* proliferum, *Moore, Ferns of Gt. Brit. Nature Printed*, t. 13 C (subvar. Wollastoni); *Id.*, *Handb. Brit. Ferns*, 3 ed. 86.
Polystichum angulare, *v.* angustatum, *Moore, Handb. Brit. Ferns*, 2 ed. 91.

Var. **cristatum**: fronds lanceolate, their apices and those of the pinnæ multifid-crisped. [Plate XXVII A.]

Polystichum angulare, *v.* cristatum, *Moore, Handb. Brit. Ferns*, 87.

Var. **polydactylum**: fronds narrow lanceolate; their apices and those of the pinnæ ramose or multifid, plane. [Plate XXVII B.]

Polystichum angulare, *v.* polydactylum, *Moore, Sim's Cat. Ferns*, 1859.

Var. **Kitsoniæ**: fronds branched above, the branches corymbosely tufted; pinnæ dilated-crisped at their apices; pinnules crowded, setaceo-serrate, more or less confluent and varied in form on the branches. [Plate XXVIII.]

Caudex thick, tufted, scaly, erect or decumbent, formed of the bases of the older fronds consolidated with a woody axis, the bases of the fronds being adherent, *i. e.*, not articulated or separating spontaneously; this caudex sometimes becomes lengthened, acquiring a trunk-like character in very luxuriant old plants. *Scales* similar to those borne on the stipes. *Fibres* numerous, strong, coarse, branched, dark-coloured, tomentose.

Vernation circinate, the main rachis becoming recurved when the frond is about half developed, the pinnæ convolute towards the main rachis.

Stipes rather lengthened, usually from four to six inches long, sometimes considerably longer, densely scaly, with long lanceolate-acuminate and linear-lanceolate scales of a reddish-tawny colour and dry membranous texture; these again intermixed with numerous others, both smaller hair-like ones and adpressed ciliated scurf-like ones, which are continued over the rachis; the larger and broader ones gradually diminishing from the base upwards. *Rachis* prominent, rounded behind, slightly channelled in front, shaggy with the numerous hair-like scales already mentioned.

Fronds from two to four or five feet high, and from seven to ten inches across at the broadest part, herbaceous or subrigid, full green above, paler beneath, usually lax, spreading and more or less arched or drooping, numerous, arranged in a circlet around the crown, lanceolate, bipinnate or tripinnate. *Pinnæ* numerous, nearly linear, rather broadest at the base, tapering towards and acuminate at the apex, alternate, sometimes distant, the basal ones usually diminishing somewhat in length, but in some varieties longest. *Pinnules* somewhat crescent-shaped, *i. e.*, ovate-falcate, with a strong anterior auricle or projecting lobe, flat, acute or bluntish, distinctly often deeply serrated on the margins, the serratures tipped with soft slender bristles, which are usually most strongly developed at the apex of the pinnule and of the auricle; the base is not toothed, but is somewhat rounded on the posterior side, truncate but with a convexity on the side parallel with the rachis, so that the base becomes an obtuse angle with slightly curving sides, at the apex of which is placed a short but distinct slender stalk diverging from the rachis at an angle of about 45°, and by this the pinnules are attached to the rachis of the pinnæ. The basal anterior pinnule is usually somewhat, often much larger than the rest, and is in some plants deeply pinnatifid or even pinnated, and occasionally other pinnules situated near the base of the pinnæ are deeply divided. The under surface of the pinnules is furnished with fine scattered hair-like scales.

Venation of the pinnules consisting of a flexuous *costa* or midvein with alternate *veins* or branches; these veins are furcately branched, producing two, three, or more *venules* in each fascicle. The anterior venule of the fascicle bears a sorus, at or very near to its apex. The auricle has a stronger vein, which is pinnately branched, producing several simple or forked venules, of which some three or four generally bear sori.

Fructification on the back of the frond, generally occupying the whole of the upper part to the extent of two-thirds, but sometimes confined on this portion to the upper part of the pinnæ. *Sori* small, numerous, round, indusiate, seated at or near the apex of the venule, forming a line on each side of the midvein, and also of the vein of the auricle, often crowded, and sometimes becoming confluent; they are attached to the anterior venules of the fascicle whenever the veins are forked, but in the auricle several of the simple venules are fertile. *Indusium* firm, membranaceous, orbicular, peltate or umbilicate. *Spore-cases* numerous, brown, roundish obovate. *Spores* roundish, ovate, muriculate.

Duration. The caudex is perennial. The fronds, moreover, are persistent, and in mild seasons and sheltered situations, the plants retain their fronds in a tolerable fresh state far into or sometimes throughout the winter. Under shelter, the species is decidedly evergreen, the old fronds only gradually yielding, when the new ones become developed, which occurs about May.

Although as regards *P. angulare* and *P. aculeatum* there is so close an affinity, that instances occur in which it is difficult to determine between them, yet, confining our view to the plants as found in Great Britain, such instances are rare, at least to those who have made themselves familiar with the aspect and characteristics of the plants. As to the application of the names, there is doubtless a certain amount of error and confusion, which it is hoped our autographic delineations may assist in correcting. When the inquiry is, however, extended, so as to include the closely allied exotic Ferns, the limits of the species become indistinct; and it is perhaps doubtful whether in this more comprehensive view,—at least by means of the mutilated examples alone available for examination in herbaria,—they can be defined with sufficient clearness to be kept permanently separate. The study of living plants may, indeed, afford other distinctive marks than those derived from form and texture, as in the case of *Polypodium Dryopteris* and its ally, which have a different vernation, and in that of some forms of *Lastrea Filix-mas*, among which differences both of vernation, and in the structure of the indusium occur. In reference to *Polystichum angulare*, there seems, with our present information, no mean between the two extremes of uniting the whole series from the simply pinnate *P.*

Lonchitis to the tripinnate *P. angulare*—an unbroken series being traceable; or, of retaining the three British species we have figured [Plates XV., XVI., and XVIII.], as well as some of the allied exotic ones, as distinct. We are not prepared to adopt the former alternative, and therefore, with all its doubt, prefer the latter.

The specific name *angulare*, which has been generally employed whenever the species has been kept distinct, is here retained for this plant, from a suspicion that it may, after all, be found necessary to merge it in *P. aculeatum*, in which case any present change would be impolitic. We have no doubt, however, that both the *P. setiferum* of Forskal (1775) and the *Polypodium appendiculatum* of Hoffmann (1795) are referrible here, and these names certainly claim priority over *angulare* (1810). The law of priority, which should take effect if our plant should finally be held distinct, would give the name of *Polystichum setiferum*, which is a remarkably suitable one.

The chief differences on which we rely to distinguish between the British *P. angulare* and *P. aculeatum*, consist in the obtuse-angled base of the stalked pinnules of the former, and the acute-angled or wedge-shaped base of the sessile pinnules of the more divided states of the latter. The less divided forms of *P. aculeatum* are much less likely to be misunderstood. Even in young immature plants of *P. angulare*, the stalked pinnules may be met with, about the base of the lower pinnæ, at an early stage of growth, so that they need not be mistaken. The upper parts of the fronds alone, in these *Polystichums* are useless for the purpose of identification.

This very beautiful Fern, which delights in shady wooded places, woods, and hedge banks, is much less frequent in its occurrence in the United Kingdom than *P. aculeatum*, though probably more common than it in the south of England and in Ireland. It appears to extend over the whole of England and Wales, in greater or less profusion; and there are records of its occurrence in Scotland. In Ireland, it is reported from all the provinces; and it moreover occurs in Jersey. It is computed by Mr. Watson to range from the coast level to an elevation of from 300 to 600 feet. The habitats of which we possess notes, are the following:—

Peninsula.—Cornwall: Penzance. Devonshire: Lynmouth; Ilfracombe, *Rev. J. M. Chanter;* Barnstaple, *F. Mules;* Ottery St. Mary, *G. B. Wollaston;* between Totness and Ashburton. Somersetshire: near Bath; Nettlecombe; Selworthy.

Channel.—Hampshire: Stubbington; Uplands; Cattisfield, Basingstoke, and elsewhere. Isle of Wight. Dorsetshire. Wiltshire. Sussex: Hastings; Cuckfield; Patching; Findon, etc.

Thames.—Hertfordshire: Panshanger; Hatfield Woodside; Colney; Watford; Totteridge. Middlesex: Brentford. Kent: Sturry; St. Mary Cray; Rainham; Tunbridge, and elsewhere. Surrey: Mayford; Stoke; St. Martha's Hill, near Guildford. Essex: Epping; Springfield.

Ouse.—Norfolk: Norwich. Huntingdonshire.

Severn.—Gloucestershire: Leigh Woods, near Bristol; Forest of Dean; Broadwell, *H. Buckley.* Warwickshire: Stoneleigh; Berkeswell; Rugby; Hearsall, etc. Herefordshire: Ross. Worcestershire: Eartham; Malvern; Suckley. Staffordshire. Shropshire: Blodwell Rocks; Wenlock.

Trent.—Derbyshire: Matlock. Leicestershire.

Mersey.—Lancashire: Chaigeley, near Clitheroe; Manchester; Prescott; Hail Wood. Cheshire.

Humber.—Yorkshire: Ingleborough; Edlington Crags, near Adwick; Roche Abbey; Mulgrave Castle, *J. Horsfal;* Halifax; Richmond; Heckfell Woods; Elland, and other parts.

Lakes.—Westmoreland: Loughrigg Fell; Ambleside. Isle of Man.

S. Wales.—Pembrokeshire: Tenby. Glamorganshire: Gower. Brecknockshire: Talgarth. Cardiganshire. Radnorshire, common.

N. Wales.—Anglesea: Beaumaris; Cickle. Carnarvonshire: Conway; Bangor. Denbighshire: Ruthin, *T. Pritchard.* Flintshire: Rhyl, *H. L. Ensor.*

E. Lowlands. — Berwickshire: Peasebridge. Edinburghshire: Corstorphine, *R. M. Stark.*

W. Lowlands.—Ayrshire: Kelburn, near Largs, in considerable quantity, *A. Tait.*

W. Highlands.—Argyleshire: Ederline, Loch Gilphead, *Mrs. A. Smith.*

Ulster.—Antrim: Blackstaff Lane; Colin Glen, Belfast.

Connaught.—Arran Isles. Galway: Connemara; Blackwater, near Gort, *J. R. Kinahan.* Leitrim: Glen Car, *R. Barrington.*

Leinster.—Wicklow: Tinnahinch, *C. C. Babington*; Newtown Mount Kennedy, *R. Barrington.* Dublin: Ballinteer, *J. R. Kinahan.* Kilkenny, *J. R. Kinahan.*

Munster.—Cork: Clonmel, *J. Sibbald.* Waterford. Tipperary. Clare, *J. R. Kinahan.* Kerry: Killarney, *R. Barrington.*

Channel Isles.—Jersey, *Dr. Allchin.* Guernsey, *C. Jackson.*

There are but scanty records of the occurrence of this Fern in the northern parts of Europe. It has been found in Sweden and Norway, according to Fries, and Nyman also records it from Gothland, Norway, and Denmark. In Scotland, as already intimated, it is rare, more plentiful in England, especially towards the south, and becoming even abundant in Ireland. It is also found plentifully in the central and southern parts of Europe,—in France, Belgium, and various parts of Germany, on the Pyrenees, in Spain, in Tuscany, Naples, and other parts of Italy, in Dalmatia, Croatia, Transylvania, and Hungary, and in Greece. We believe it is also met with at the Dardanelles, and on the coast of the Black Sea, for we have no doubt, from Forskal's brief specific phrase, that the *Polypodium setiferum* from the former habitat is the present species rather than *P. aculeatum* under which it is usually quoted; and specimens of D'Urville's gathered on the shore of the Black Sea, preserved under the name of *aculeatum* in the Museum of Natural History at Paris, of which we have seen a sketch, belong to the acute-pinnuled forms of *P. angulare.* In Asia, it would appear to have been found in Georgia, and in the Province of Guriel at the eastern extremity of the Black Sea; and thence it extends to India, where it is found in Kashmir, Simla, Khasya, Kumaon, in the valley of the Indus at Balti, in Nepal, at Madras, and in the island of Ceylon. In Africa, a Fern quite accordant with the British plant, occurs in the Canary Islands, Madeira, and the Azores; also, in Abyssinia, on the African coast of the Mediterranean according to Kunze, and at Natal. Dr. Asa Gray reports it as occurring in various parts of the United States of America; it is found in New England, and has been gathered at Sitka; it is an inhabitant of Quito and

Chacapoyas, and of Chili; and forms almost, if not quite identical, occur in Guatemala, in Mexico, New Granada, and Caraccas. Singapore and Java in the East, yield cognate forms barely separable on the one hand from this species, or on the other from *P. aculeatum*, and similar forms appear common over India. In some of these tropical forms, the harsh texture of the European *aculeatum* is found associated in the same individual with the outline and aspect of the European *angulare*, rendering it difficult, if not impracticable, to point out the limits of the two species.

In the garden, this Fern will be found very ornamental, and of very easy management. It is in fact, one of the most beautiful of our hardy species, its value being enhanced by its evergreen character, and by its capacity for submitting itself to the vicissitudes of artificial cultivation. It grows readily in free sandy loam, either in shady parts of the garden or shrubbery, on out-door shady rock-work, or in pots in the in-door Fernery, in the latter case requiring a tolerable amount of pot-room, and then attaining remarkable elegance. It is increased with facility by division whenever lateral crowns are produced.

A remarkable proliferous or viviparous character has been observed in several of the varieties of this species, as well as in many other British Ferns, including *Polystichum Lonchitis*; *P. aculeatum*, with its variety *lobatum*; *Lastrea Filix-mas*, two varieties; *L. æmula*; *Asplenium lanceolatum*; *A. Ruta-muraria*; *Scolopendrium vulgare*, several varieties; *Blechnum Spicant*, etc. Some of the varieties of the present species propagate extensively by means of these bulbils, which form either towards the base of the stipes, or along the rachis in the axils of the lower pinnæ, or in some instances on the veins of the fronds. Although among exotic Ferns instances of this viviparous growth were known to occur frequently, yet our acquaintance with so many bulbil-bearing British Ferns is due to the scrutiny of a few zealous cultivators, especially Mr. Wollaston of Chislehurst, Dr. Allchin of Bayswater, Mr. Clapham of Scarborough, and Mr. Baxter of Oxford. Most of the instances above referred to were observed during the summer of 1854. Mr. Baxter has suggested that it may be a result of pot-culture, all the instances in which it has been observed, having been on potted plants. We think it may be the combined result of the check caused by the cramping of the roots incidental to pot-culture, and the excitement arising from the very moist atmosphere which is kept up in most Fern-houses. The instances thus observed, however produced, appear to afford additional evidence that the fronds of Ferns are not leaves, as some would call them, to which, however, the fact of their normally bearing the fructification seems repugnant; but that they at least include something of the nature of branches. Another fact may be mentioned as militating against the opinion that the fronds of Ferns are mere leaves. Leaves, it is maintained by physiological botanists, have their points first formed, the perfected apex being as it were pushed forward by accretion from below. Now in the fronds of Ferns, it may often be seen to demonstration, that the lower parts are perfectly developed and bear mature sori, whilst the apex is still unrolling; this is very obvious in the genus *Nephrolepis*. Besides the bulbilliform mode of increase above adverted to, Mr. Wollaston has observed a different kind of what is supposed to be viviparous development in the *Polypodium vulgare* var. *omnilacerum*, on a plant communicated by Mr. E. T. Bennett. In this case, the development consisted of prothalloid growths, on the apices of the serratures of the lobes; these had every indication of being capable of further evolution, though unfortunately the frond was broken off before they were observed, so that their vital energy could not be fully tested.

Polystichum angulare is one of those Ferns which exhibit a very large degree of variation, the differences in some instances being very marked. There are several distinct modes of variation. The blunt-pinnuled broad-fronded form is taken as the type. Diverging from this, some forms have the fronds small and extremely narrow; others have the pinnules narrow and acute; while in some they are very densely placed, and in others they are distant. One remarkable form has the pinnules nearly round and entire. Some have the secondary rachides distinctly winged, and uniting the bases of the pinnules into a confluent mass, thus obscuring one of the essential characteristics of the species—the stalked pinnules. A set of forms have the margins biserrate or lobato-serrate. Other forms are variously depauperated, giving rise to several very handsome variations which are found to retain their peculiarities. Some forms are thin in texture, pale-coloured, and much divided, having a plumy appearance; and not a few different forms are now known in which cristate tasselled terminal tufts are developed in different ways. The plant being evergreen, and extremely varied, may perhaps be entitled to rank as the most beautiful among our British species.

1. *angustifrons* (M.). A very interesting form, remarkable on account of the narrowness of its dwarf attenuated fronds, which are in the specimens before us, about eight inches long, exclusive of the stipes, and barely more than an inch in width, the upper third being fertile. Fronds narrow linear-lanceolate, attenuated at the apex, distinctly bipinnate; the pinnules are small, close-set, normal in character, bristly-serrate, and auricled: about three pairs on the larger pinnæ being stalked, the rest confluent. It was found at Barnstaple, Devonshire, by Mr. Jackson. Mr. Clapham reports it to be sometimes cornute.

We have another small and narrow fertile form of this species apparently differing from *angustifrons*, provisionally named *stenophyllum*. The fronds are nearly nine inches long, and almost two inches across the middle, the upper part being attenuated; they are bipinnate, the lower pinnules distant, the basal pair only stalked and auricled, and the rest confluent with a rounded anterior margin, in this respect closely resembling *P. aculeatum*, v. *lobatum*. It was gathered by Mr. Wollaston, in Devonshire.

2. *hastulatum* (M.). This is a small pinnuled form, quite like the Italian *hastulatum* of Tenore, as figured in the *Flora Napolitana*. It is chiefly remarkable for the small size of the acute pinnules, and for their distinct and slender footstalk; the auricle, too, is very distinct, acute, and in the case of the lower pinnules is separated by a deep incision from the rest of the pinnule. It has been found in—Surrey: St. Martha's Hill, near Guildford. Devonshire, *Rev. J. M. Chanter*. [Plate XIX D.—Folio ed. t. XII B.]

3. *quadratum* (M.). A remarkably neat small-pinnuled form, the chief peculiarity of which is the approach to squareness in the

outline of the little trapeziform pinnules. In all other respects it is normal. The pinnules are sharply, but simply toothed. The prettiest forms we have seen are from—Devonshire, *G. B. Wollaston.* Somersetshire: Nettlecombe, *C. Elworthy.* Yorkshire: Whitby, *W. Willison.* These have fronds about a foot high with a caudately attenuated apex. Other analogous forms are from—South Devonshire, *Miss A. Hoseason.* Yorkshire: Black Moor near Helmsley, *A. Clapham.* Antrim, Ireland, *A. Stansfield.*

4. *affine* (M.). An elegant form with the aspect of a slender *P. aculeatum.* Mr. Wollaston regards it as a distinct species, and calls it *P. affine.* The fronds are long-stiped, dark-coloured, firm, lanceolate, attenuated at the apex. The pinnules are ovate-falcate, blunt, bluntly auricled, rather convex, finely and indistinctly serrated. It was found at Hartley in Hampshire, by Mr. Wollaston.

5. *acutum* (Woll.). This form, which in general aspect is normal, has the pinnules acute, and very distinctly stalked, more elongated than in *hastulatum* (2), rather narrow, often falcate, and strongly auricled; the larger forms are lobate-serrate, but sometimes the serratures are only slightly developed, the apices of both the pinnules and auricles being however aristate. It has been found in—Sussex and Hampshire, by Mr. Wollaston. Lancashire: Moorhills Clough, near Bromley, *A. Stansfield.* A thin-textured form (*acutum dissectum*), having the same narrow elongated acute pinnules, but more lobate, and growing to a considerable size, has been found in—Westmoreland: Whitbarrow, *F. Clowes.* Somersetshire: Nettlecombe, *C. Elworthy.* Devonshire: Barnstaple, *C. Jackson.*

6. *aristatum* (Woll.). This, although not unlike the normal state of the species in its form and habit, differs in having the points of the serratures much more distinctly aristate, the long hair-like points standing forwards in a remarkable way, giving the plant a bristly appearance. The original variety found by Mr. Wollaston in Sussex, is, moreover, proliferous, producing its bulbils on the stipes, either beneath or at the surface of the ground. Another remarkable aristate form has been sent from Shebden Dale, near Halifax, by Mr. Stansfield.

7. *incisum* (Woll.). This variety, of which there are two forms, is a large-growing plant. The pinnules are very dissimilar in size

and shape; a few on the lower portion of the frond, are divided as in *subtripinnatum*, but as they approach the apex they become less divided, and are variously incised and irregularly laciniated or jagged, some here and there being depauperated; their segments are again serrated, the lower anterior ones being prolonged into an auricle. It was found by Mr. Wollaston, at Littlehampton, in Sussex. Another form found by Dr. Allchin in Sussex, is of smaller growth, and is occasionally marked by a disruption of the epidermis on some of the pinnules.

8. *latum* (M.). A variety with short broad ovate pinnules, resembling the continental *Braunii*. Fronds lanceolate, bipinnate; pinnæ bluntish, sometimes with a cristate tendency; pinnules broad ovate with the anterior base enlarged, setosely-serrate, or when large lobate-serrate. Devonshire: Barnstaple, *C. Jackson;* Ilfracombe, *J. Dodds.* Somersetshire: Nettlecombe, *C. Elworthy.*

9. *densum* (M.). This variety is of suberect habit, and is remarkable from the rachides and veins being densely clothed with hair-scales, while the stipes is thickly coated with larger and broader scales. The fronds are bipinnate, oblong-lanceolate, the pinnæ rather unequal in length. The pinnules are small, crowded, oblique oblong, bluntish, with a large distinctly separated obovate auriculiform lobe, and the rest of the margin lobate-serrate, with aristate teeth. It is an elegant form. The plant was found at Albury, Surrey, by Mr. Morse. [Plate XX C: from a young plant.]

A dense-pinnuled form (*stipatum*, Woll.) has been found by Mr. A. Tait. This is exactly lanceolate, tapering both above and below; the pinnæ short, overlapping; the pinnules also crowded and overlapping, ovate, acute, auricled, serrate, the basal one only having a very short stalk, the rest being decurrent.

10. *imbricatum* (M.). This graceful and curious variety is perfectly distinct from all others. The fronds are about two feet high, linear-lanceolate in outline, attenuate at the apex. In their narrow form, as well as in habit, they resemble the *Polystichum lobatum* of authors, though they have all the important characters of *P. angulare.* The pinnæ are short, linear-oblong, bluntish at their apices, spreading. The pinnules are crowded and imbricated, roundish-oblong, auricled at the anterior base, scarcely narrowed

at the apex, obscurely serrate, but terminating in a sharp mucronate or setiferous tooth, the auricle also being tipped by a similar tooth; the basal anterior pinnule is larger, lobate-serrate, distinctly stalked, while the upper ones are connected with the rachis by a short somewhat winged decurrent petiole. This remarkable variety is one of those which possess the property of producing bulbils or gemmæ on the stipes of the larger fronds. It was found near Nettlecombe, in Somersetshire, by Mr. Elworthy, gardener to Sir W. C. Trevelyan, Bart. [Plate XXI.—Folio ed. t. XII E.]

11. *rotundatum* (M.). A very remarkable form. The fronds are narrow lanceolate; the pinnæ short, and terminating in a blunt confluent lobe; the pinnules roundish oblong, or often nearly round, quite obtuse, the margins obscurely crenate, not at all spinulose or setaceous. The fronds are rather small. It has been only recently found, by Mr. Elworthy, near Nettlecombe, in Somersetshire.

12. *decurrens* (M.). This is a very distinct and handsome variety. The fronds are ovate or lanceolate, bipinnate, rather thick in texture, sometimes multifid at the apex. The lower pinnæ are the more normal in character, while the upper ones, which are fertile, are more or less contracted and altered in form. The pinnules are distant, those of the lower pinnæ oblong acute, with a large anterior auricle, wedge-shaped at the base, and nearly all of them decurrent with the rachis; they are deeply and rather distantly lobate-serrate with spiny serratures. The pinnules of the upper pinnæ are more decidedly decurrent, smaller, more distinct, generally with the auricle developed, but the remaining part of the pinnule much reduced in size. It is a native of Somersetshire, and was found at Nettlecombe by Mr. Elworthy.

13. *alatum* (M.). This is a most remarkable variety. Its peculiarity consists in the pinnules being all connected by a very obvious wing which borders the little footstalks, as well as the secondary rachides, on which they are thus decurrent. The fronds grow a couple of feet high, and are broadly lanceolate, bipinnate. The pinnules are short, but acute, with the anterior basal lobe or auricle much developed; and the margin is divided into shallow rounded lobes, which are indistinctly serrate, but tipped by a bristle; the under surface is also densely covered with hair-like scales. It was

found at Selworthy, in Somersetshire, by Mrs. Archer Thompson, and seems unique. [Plate XXII.—Folio ed. t. X C.]

14. *pterophorum* (M.) A form analogous to *alatum* (13), agreeing in the remarkable peculiarity of having the secondary rachides winged, and the pinnules consequently united, but differing in being a less scaly plant, in having the pinnules less lobed, in the serratures being appressed, and in the auricle being almost wanting. It was found in Devonshire, near Ottery St. Mary, by Mr. Wollaston. [Plate XX A.]

15. *biserratum* (M.). This is one of the large growing forms, and is remarkable for its large broad pinnules, which are stalked, inciso-serrate, the basal anterior lobe separated by a deeper incision, the rest biserrate and in most cases conspicuously aristate. It frequently has very long stipites. It seems to be not uncommon, since specimens which we refer to this form have been obtained from—South Devon, *Miss A. Hoseason;* Ilfracombe, *Rev. J. M. Chanter;* Barnstaple, *H. F. Dempster*, *C. Jackson.* Somersetshire: Nettlecombe, *C. Elworthy.* Sussex: Littlehampton, *G. B. Wollaston.* Middlesex: Osterley Park, and Lampton Lane near Brentford, *S. F. Gray.* Gloucestershire: Broadwell, *H. Buckley.* Yorkshire, Whitby, *W. Willison;* Mulgrave Castle, *J. Horsfal.* Pembrokeshire: Castle Malgwyn, near Llechryd, *W. Hutchison.* Dublin: Glendruid, *R. Barrington.* Jersey, *Dr. Allchin, C. Jackson.* [Plate XIX C. —Folio ed. t. XII D.]

Mr. Jackson has found near Barnstaple, Devonshire, three forms which we may mention here:—(*a*) a very large form (*magnum*), the pinnæ 6½ inches long, pinnules ¾ of an inch, lobate and biserrate; (*b*) another large form with shorter pinnæ than the foregoing, but with pinnules equally large, deeply lobate and biserrate, the tips of the lobes terminating in a very long seta (*setigerum*); (*c*) pinnules oblique-oblong obtusely rounded (*obtusum*), unequally spiny-serrate, scarcely auricled or lobed. They come near *biserratum.*

16. *latipes* (M.). This form, related to *biserratum* (15), has been found at Nettlecombe, Somersetshire, by Mr. Elworthy. It is a large plant, the pinnules rather more elongated and acute, as well as more deeply lobed than in *biserratum*, but its chief peculiarity consists in the lowest pinnæ being considerably the longest; the base of the frond measures upwards of ten inches across.

17. *intermedium* (Woll.). This is a thick robust, fleshy-looking, rigid, upright plant, in these respects resembling *P. aculeatum.* The fronds are ovate-lanceolate, large, distinctly bipinnate. The pinnules are crowded and often overlapping; subtrapeziform from the development of the anterior basal angle; deeply inciso-serrate along the margin, the basal anterior lobe being much enlarged, and all the segments biserrate, with the teeth aristate. The fronds are occasionally multifid, and not unfrequently abrupt at the apex, and when so have a tendency to produce bulbils. It has been found in —Kent: near St. Mary's Cray, *R. Sim.* Sussex: Newick Park, *J. H. Slater;* Littlehampton, *G. B. Wollaston.* Somersetshire: Nettlecombe, *C. Elworthy.* Anglesea, *T. Pritchard.* Glamorganshire: Swansea. Pembrokeshire: Castle Malgwyn, near Llechryd, *W. Hutchison.* Guernsey, *C. Jackson.* [Plate XIX B.]

18. *trapezoideum* (M.). This is a dwarf form, somewhat in the way of *intermedium, i. e.,* with short broad biserrate pinnules, with the anterior angle much developed, diminishing, and becoming somewhat confluent in the upper part of the frond. The fronds are narrow ovate attenuate, bipinnate; pinnules broad, almost trapeziform, biserrate, confluent near the tips of the pinnæ; those of the upper pinnæ somewhat smaller, and much more confluent. Devonshire: Ilfracombe, *Rev. J. M. Chanter.* Somersetshire: Nettlecombe, *C. Elworthy.* Lancashire, *R. Morris.* Sussex: Littlehampton, *G. B. Wollaston.* Kent: Sturry. Kerry: Killarney, *R. Barrington.*

19. *pulchrum* (M.). The fronds of this form are lanceolate; the pinnules ovate, strongly auricled, inciso-lobate, with unequally serrated lobes; the fertile pinnules in the upper part of the frond are rather more deeply lobed, smaller, somewhat decurrent; here and there a few pinnules are depauperated. Somersetshire: Nettlecombe, *C. Elworthy.*

20. *irregulare* (M.). This is a very curious form, large and ovate-acuminate in outline, bipinnate. The lower pinnæ, which are the most perfect and normal in character, are longest and bear varying and unequally inciso-lobate pinnules, the basal anterior lobe of which forming the auricle, is much enlarged, and considerably detached, while the rest form lacerate serratures or lobes, all the larger of which are again serrated. The upper pinnæ are fertile, more or

less depauperated, distant, decurrent, and irregular in size, outline, and toothing. It was found by Mr. Elworthy, near Nettlecombe, in Somersetshire. [Plate XX D.—Folio ed. t. XII C.]

21. *grandidens* (M.). A very graceful variety. The fronds are of moderate size, narrowly but irregularly lanceolate, the pinnæ being of various lengths, though less depauperated than in either *interruptum* (22), or *dissimile* (23), which are nearly related forms. The pinnæ very frequently terminate abruptly, either in a fan-shaped confluent leaflet, or in a leaflet resembling the ordinary pinnules. The pinnules are of different shapes—ovate, oblong, roundish, or obliquely wedge-shaped, and are very remarkable from the large coarse deeply-cut teeth of the margin (laciniate-dentate), which are acuminately setaceous. When fertile the pinnules in the upper part of the frond often grow out into irregular sharp-pointed divergent angular lobes. The most marked forms of the variety produce very narrow fronds, the pinnæ being all and nearly equally abbreviated, the pinnules small, wedge-shaped, with a few coarse angular teeth, and the terminal one larger and fan-shaped. Sometimes the pinnules are all larger but of the same laciniate-dentate character. The best forms we have seen were found in—Devonshire: Totnes, *R. Penwill.* Somersetshire: Nettlecombe, *C. Elworthy*—two forms, one dwarfer than the other. Ireland: Killarney, *S. Jervis.* Fermanagh: Lisnaskea, *Rev. W. R. Bailey.* Other slightly varying forms come from—Devonshire: Totnes; Barnstaple, *C. Jackson.* Yorkshire: Rolston Scar, *R. Foxton*—rather more unequal in the pinnæ. Fermanagh, *Rev. R. Eccles*; banks of Lough Erin, *Rev. W. R. Bailey.* Antrim. [Plate XX B, and XXV A.]

22. *interruptum* (Woll.). This is a variety of remarkably irregular growth. In its characteristic state the majority of the pinnæ are greatly abbreviated and more or less truncate or abrupt, sometimes reduced to the size of ordinary pinnules, while here and there one of full length is produced. These longer pinnæ usually have a portion of their pinnules normal, or nearly so, in character, oblong and biserrate, while others are reduced to half their size, and roundish or ovate in outline without auricles, and wedge-shaped at the base. The abbreviated pinnæ have their pinnules changed to a roundish or ovate form, with incised or laciniated margins. The apex of the

frond is caudate. It is similar in general character to *dissimile*, but differs in the more elongated form and less setaceous teeth of those pinnules which approach the normal character. It has been found in—Hampshire: Fordingbridge, *G. B. Wollaston*. Devonshire: Ilfracombe, *Rev. J. M. Chanter*—the pinnules generally more deeply laciniate-toothed. Killarney: *Dr. Kinahan*—rather less toothed in the normal portions. Guernsey: *J. James*.

23. *dissimile* (M.). This plant in its normal parts resembles *intermedium* (17), having its pinnules ovate, and their segments biserrate and more aristate than usual, and it is also proliferous; but it differs from that variety in having its fronds frequently much depauperated, the pinnæ variously abbreviated, irregularly truncate, or multifid, often here and there intermixed with one of normal outline; the pinnules also are very irregular in size and shape, often fan-shaped or depauperately wedge-shaped. It has, moreover, a hispid aspect, arising from the abundant aristate tips of its serratures, and the narrowness of the scales with which it is profusely covered. It was found near Tunbridge Wells, Kent, by Mrs. Delves, and two or three similar forms have been gathered by Mr. Jackson, at Barnstaple, Devonshire, and by Mr. James, in Guernsey. [Plate XXIV.]

24. *supralineatum* (M.). This variety, which requires further testing, is bipinnate and apparently dwarfish. It has the pinnules more or less and variously depauperated, and variously incised, lobed, or toothed, the teeth being not unfrequently rounded, and generally wanting the usual seta. The peculiar feature of the variety is, however, the presence on the upper side of the pinnules of an excurrent membrane, sometimes forming a rib-like line, sometimes running out into callous teeth. It was found near Barnstaple, Devonshire, by Mr. C. Jackson.

25. *inæquale* (M.). This is rather a neat-looking abnormal form, the pinnules of which are not much altered. The lower pinnæ are short, giving the frond a lance-shaped outline. The majority of the pinnules are oblong-ovate, bluntish, with a terminal awn and an anterior basal auricle, as in the neat normal forms, but they are here various in size, some being broader, some narrower, and, in the upper part of the frond especially, variously misshapen and to some extent depauperated. It is the irregularity in the size and form of

the pinnules which is the most remarkable feature in this variety, which is of average size, and fertile towards the apex. It was found in Devonshire by Mr. Wollaston.

26. *præmorsum* (Allch.). This is a curious rather than a handsome variety. The fronds are narrow, dwarfish, bipinnate; the pinnæ short, with only a few pinnules distinct, the rest confluent or congested into a variously lobed or laciniated spiny-toothed, often truncate præmorse or partially depauperated apex. The few pinnules are rather large, oblong, and biserrate in a crowded but irregular manner. The upper pinnæ are entirely confluent into misshapen pinnæ, variously pinnatifid or laciniated, the lobes unequally biserrate, the teeth being frequently aristate. The pinnules or lobes are very frequently marginate, that is, having a distinct excurrent membrane, here and there apparent, as in some forms of the Hart's Tongue Fern. This variety was originally found in Ireland by Dr. Allchin, and was very imperfectly described in our folio edition. More perfect specimens found also in Ireland, Tyrone: Clogher, by the Rev. W. R. Bailey, enable us to notice it more fully. It has besides been found at Barnstaple, Devonshire, by Mr. C. Jackson. A form of similar character here and there interrupted, but not misshapen as in true *præmorsum*, has been found at Oldstead in Yorkshire by Mr. C. Monkman.

27. *gracile* (Woll.). A truly graceful variety, having the general characteristics of *lineare* (28), but distinct. The frond is ovate attenuate at the apex, bipinnate; the pinnules smallish narrow and distinct, and the whole habit lax. These pinnules are rather irregular in form, but generally narrow oblong acute, scarcely auricled, with the margin inciso-serrate rather than lobate; many of the pinnules are linear with a wedge-shaped base. The pinnules on the upper pinnæ become very acute, and bear each three or four distinct sori. The apices of the lower pinnæ, as well as the entire upper pinnæ, are confluent, lobate-serrate, and the apex of the frond itself is lengthened and slender so as to become caudate. It was found in Devonshire by Mr. Wollaston. [Plate XXV B.]

28. *lineare* (M.). This may be regarded as a more decidedly bipinnate form of *confluens* (29); it is moreover much larger, and is probably distinct. The plants are tolerably constant, producing in

some instances only an occasional characteristic frond, in others yielding them more freely. It is rather lax in habit. When well marked, the lower half of the pinnæ are furnished with tolerably perfect pinnules, which are linear and only here and there and then strongly auricled; then come a few abortive pinnules, and finally a confluent linear lobate-serrate apex. The upper third of the frond resembles a whole frond of *confluens* (29), the uppermost pinnæ being entirely confluent as in that form. The pinnules are here and there and very irregularly depauperated, and the large anterior basal pinnule is by no means constant. In some fronds the whole of these basal pinnules are excessively depauperated, so as to be all but entirely wanting. The more normal-looking fronds occasionally produced, sometimes have the pinnules ovate auriculate, and toothed in the way of *biserratum*, or, sometimes narrowly oblong with a large auricle, approaching nearer to the characteristic state. Mr. James, by whom the variety has been cultivated in the mild climate of Guernsey, remarks, that of the two kinds of fronds, the narrow or characteristic ones are most numerous, while the broader ones are of autumn growth. In cases like this, it not unfrequently happens that the characteristic development is checked by division or by repotting or by some disturbance of the root, or of the conditions of growth, and gradually returns as the plants become established. This variety is of Devonshire origin, and was found by Mr. C. Jackson. [Plate XX E.]

29. *confluens* (M.). A curious dwarf semidepauperated variety, but symmetrical, the pinnæ being uniformly affected. The fronds are ovate-lanceolate, attenuated at the apex, scarcely bipinnate, the lower anterior pinnules only being developed, and these forming a conspicuous row on each side the rachis; beyond this the pinnules are more or less depauperated, often wedge-shaped, subauriculate, or aristate, and soon becoming confluent into a linear lobate-serrated apex. The upper pinnæ are entirely confluent, becoming linear-falcate in outline, with a strong anterior auricle, and serrated. It was found in Ireland by Mr. S. Foot, and was communicated by Mr. D. Moore. Mr. Elworthy has found near Nettlecombe in Somersetshire, a somewhat similar variety; and another analogous form has been found at Torquay. [Plate XXVI.]

30. *dubium* (Woll.). This is a thick harsh form, more rigid than usual, with much of the aspect of *P. aculeatum,* having also an enlarged anterior basal pinnule as in that species. The form and attachment of the lower pinnules, however, unites it with the present species, of which it may be regarded as a form approaching *P. aculeatum.* Sussex, *G. B. Wollaston.* We regard the following as being the same form:—Hampshire: Basingstoke, *F. Y. Brocas.* Wiltshire: Salisbury, *W. Moore.* Somersetshire: Nettlecombe, *C. Elworthy.* Lancashire: Wood Plumpton, *T. Stansfield.* Denbighshire: Rhuabon, *A. L. Taylor.* Antrim, Ireland, *D. Moore.*

31. *plumosum* (Woll.). This is one of the most beautiful of the varieties of this charming fern. The fronds are rather large, pale-green, broad, having an ovate-lanceolate outline, bipinnate, becoming almost tripinnate in the most divided parts, thin and dry in texture, becoming papery when dry. The pinnules are long-stalked, deeply inciso-lobate, and give a feathery aspect to the gracefully arching fronds, which well deserve the name *plumosum.* The thin texture and deep cutting of the parts are the most important characteristics. The basal anterior lobe of the pinnules is large, forming the usual auricle, and this is lobed on the margin, or biserrate with sharp teeth, the other parts being deeply incised, each lobe directed forwards, and again cut into sharp-pointed teeth. The rachis of the pinnæ is very slender. It was found in—Devonshire: near Ottery St. Mary, *G. B. Wollaston;* Barnstaple, *C. Jackson.* Somersetshire: Nettlecombe, *C. Elworthy.* [Plate XX F.]

Mr. Jackson has also found, near Barnstaple, a form evidently allied to this, and probably an accidental variation of it, in which the lobes on the posterior side of the posterior pinnules are separated into distinct linear toothed pinnulets; the pinnules on the lower side of the rachis being in fact deorsely-pinnate (*deorso-pinnatum*). It is an extremely elegant variation, but we fear not permanently distinct from *plumosum;* it requires further trial.

32. *ornatum* (M.). This variety is very ornate in character. The frond is of the normal outline, as also are the pinnæ. The pinnules are rather crowded, ovate, with a very large anterior auricle; deeply lobed, the lobes spreading and having rather open or distinctly marked sinuses, distinctly biserrate, and the auricle

lobate, with the shallow lobules biserrate. It is a short-pinnuled, though deeply-lobed form, and hence has a distinct aspect. It was found at Barnstaple, in Devonshire, by Mr. C. Jackson.

33. *tripinnatum* (M.). This very peculiar variety, which is stout, dense, and rigid in growth, has its one basal anterior pinnule on each pinna much enlarged, and much more distinctly pinnate than in other forms, though the plant is, on the whole, less divided than *proliferum* (36,) or *decompositum* (35). The most remarkable peculiarity is the unusual elongation of the anterior basal pinnules, which are nearly twice as long as the rest, and truly pinnate, the little pinnulets along nearly their whole length being distinctly stalked. The other pinnules are highly developed though less so than the basal one, and they are crowded and imbricated, and bear sori profusely. It was found at Penzance in Cornwall by Mr. Millet, and was communicated by Mr. Lowe. [Plate XIX E.—Folio ed. t. XIII B.]

34. *subtripinnatum* (M.). This is one of the more highly developed of the normal states of the species, being only somewhat less divided than *decompositum* (35). It is of large growth, and very handsome, as all the larger forms are. The lowermost pinnules, the basal ones in particular, are here so deeply pinnatifid, that the segments become almost or quite distinct; in other respects it resembles the normal type; it varies, however, in being sometimes bluntish, sometimes more acute, in the pinnules. It is reported to be common in Ireland; *e. g.*, Wicklow: Tinnahinch, *R. Barrington*; The Dingle, *R. B.*; Glen of the Downs, *R. B.* Donegal, *R. B.* Dublin: Carrickmines, *R. B.* It is, we believe, also plentiful in England in damp shady situations; we have specimens from Cornwall: Penryn, *G. Dawson*. Devonshire: Torquay, *J. Carton*. Somersetshire, *T. E. Partridge*. Sussex: Hurst Green, *Rev. J. Hand*. Kent: Sturry. Surrey: Mayford; St. Martha's Hill, near Guildford. Berkshire: Lambourne, *Mr. Walker*. Nottingham, *E. J. Lowe*. Pembrokeshire: Castle Malgwyn, *W. Hutchison*. Guernsey, *C. Jackson*. [Plate XIX A.—Folio ed. t. XIII A.]

35. *decompositum* (M.). This beautiful variety is the most compound or divided form of what may be called the normal race of the species, that is to say, those in which the general form is not much altered. The fronds are tripinnate at the base of the pinnæ,

the pinnulets being again serrated. The lowermost pinnules of each pinna are thus tolerably exact miniatures of the pinnæ of *P. aculeatum*, having their first pinnulets distinct auricled and serrated as are the pinnules in that species, while the upper ones, more or less confluent, closely resemble *P. lobatum*. The pinnules are generally acute and spinosely serrate. The variety has been observed from—Antrim, Ireland, *D. Moore*. Cornwall: Penryn, *G. Dawson*. Devonshire: Ottery St. Mary, *G. B. Wollaston*. Somersetshire: Nettlecombe, *C. Elworthy*. Kent: Tunbridge Wells, *Mrs. Delves*. Guernsey, *C. Jackson*.

Mr. Jackson has found at Barnstaple a slender divided form, which may be mentioned here, under the provisional name of *athyrioides*, given to it by Mr. Wollaston. The pinnules are narrow, almost linear, deeply lobed, with a distinct anterior auricle. The lowest pinnæ of the frond are longest, and the pinnules on the posterior side of these are much larger and more divided than those on the anterior side. It requires further trial.

36. *proliferum* (Woll.). This is perhaps the most beautiful, as it is certainly one of the most distinct varieties as yet known. Two forms have been discovered. One of these, that which has been longest known to cultivators, from having been distributed many years ago from Kew, was reported to have been found at Wimbledon, Surrey, by Mr. Choules, but of this we have specimens from Mr. Pamplin labelled Devonshire, on the authority of Mr. Choules, and suspect this to be the real habitat. The other [subvar. *Wollastoni*] was found more recently near Ottery St. Mary, in Devonshire, by Mr. Wollaston, and is a more lax and elegant plant; and the same has subsequently been found at Barnstaple by Mr. Jackson. The fronds are proliferous, bearing, chiefly at the point of junction of the pinnæ with the rachis, but sometimes in the axils of the pinnules, small bulbils from which young plants are readily obtained. Owing to its evergreen character, and its lax graceful habit, this is one of the most beautiful of hardy Ferns. When perfectly developed, especially in the case of the subvar. *Wollastoni*, the broad fronds are large tripinnate, drooping; the pinnules are narrowed and semi-depauperated, yet not distorted, but attenuated, very conspicuously stalked, distantly lobed, the

lobes being deeply divided, sometimes widely separated, so as to become pinnulets. The less elegant, because less divided, form is still remarkable for its narrowed and acute pinnules; and hence this form was called *angustatum* in the *Handbook of British Ferns* (2 ed.), but for the sake of securing as far as practicable uniformity of nomenclature for correspondent varieties of different species, the more characteristic name of *proliferum* which has been suggested, is now adopted. These forms, though less fertile than the normal state of the species, yet both bear sori. Our plate represents only a very small frond, and a pinna of a larger one. Mr. Clapham has sent us some curious sporting pinnæ of this variety, in which the parts are variously depauperated, the development in some instances resembling *grandidens*. [Plate XXIII = subvar. *Wollastoni*.—Folio ed. t. XIII C.]

37. *variabile* (M.). A large growing form with ramose pinnæ. The branching of the pinnæ is very unequal, some being divided near their base, others only at the apex, while some are unbranched, and here and there the pinnæ, whether branched or not, are abbreviated. The pinnules are slightly depauperated here and there, and where normal, are largish, acute and toothed like *biserratum*. Barnstaple, *C. Jackson*.

38. *furcatum* (M.). This is a small neat variety of which two or three states have been found in the west of England. The fronds are in great measure normal; their peculiarity consists in the tip of the frond, or the tips of the pinnæ, or both, being split or forked once or twice beyond the usual acuminated or attenuated portion; the lobes resulting from this furcation are small and short, being confined to the extreme narrowed tip. Devonshire: Gittisham, near Ottery St. Mary, *G. B. Wollaston;* Barnstaple, *H. F. Dempster*. Somersetshire: Nettlecombe, *C. Elworthy*.

39. *polydactylum* (M.). This is a very elegant plant. The fronds are slender, narrow lance-shaped, the pinnæ short and rather irregular, and the pinnules here and there abortive or depauperated, but not so much so as to affect the general outline of the frond. The pinnules when perfect are small with a very distinct stalk and auricle. The peculiar feature of this variety is the branching of the pinnæ, which become ramose at about half their length, the

branches being divergent but plane, and formed of pinnules which are in various degrees confluent. It is hence analogous to the 'many-fingered' variety of the Lady Fern. The variety is of Irish origin, having been found in 1857 in Tipperary, whence it was sent to Mr. R. Sim of Footscray. The subsequent growths have proved constant. [Plate XXVII B.]

A curious form, which may be mentioned here under the provisional name of *congestum*, has the pinnules in the upper half of the frond more or less diminished or depauperated along the basal portion of their rachis, the pinnæ being truncate, with the two or three apical pinnules acute, deeply toothed, and rather enlarged so as to form on each an abrupt terminal tuft or head. The apex of the frond, which is attenuated, is similarly affected. The lower pinnæ are normal, with small blunt pinnules. Barnstaple, *C. Jackson*.

40. *cristatum* (M.). This very beautiful variety resembles in its general features the well-known and universally admired tasselled varieties of the Male Fern and the Lady Fern; that is to say, the apex of the frond itself, as well as the apices of the pinnæ, are expanded into multifid curly or crispy tufts, those of the pinnæ being less developed than that which terminates the frond. The remaining portions of the frond are normal in character. This form has been found in several places, and there is a slight diversity among the plants, but they all agree in their general aspect and characters so closely, that they must be regarded as one form. Our figure representing only a small frond, does not show the tufts so fully developed as they often occur when in a more vigorous state; in any form it is, however, a very handsome plant. It has been found in—Devonshire: near Bristol, *J. Hillman;* Ottery St. Mary, *G. B. Wollaston;* Bickington, *C. Jackson;* and elsewhere in the same county by the *Rev. J. M. Chanter*. Somersetshire: Nettlecombe, *C. Elworthy*. [Plate XXVII A.]

41. *multifidum* (Woll.). The apex of the frond in this variety is variously branched or tufted, and various states of it have been met with, some of which prove inconstant, while others are permanent. We have received fronds from—Devonshire: Newton Abbott, *W. Green;* Ilfracombe, *Rev. J. M. Chanter;* Barnstaple, and Westleigh, *C. Jackson*. A multifid form referrible to the

var. *acutum*, has been found at Upcott, Devon, by *Mr. T. Wren Harding*.

42. *Kitsoniæ* (M.). This beautiful variety is remarkable from the rachis of its fronds separating towards the top, into four or five branches, the branches being corymbosely-tufted, and the pinnæ which form them, dilated and crispy at their tips. The lower pinnæ are normal in character, their pinnules numerous, oblong-acute, setaceo-serrate, and less auriculate than usual, the confluent tips of the pinnæ having a tendency to dilatation. In the branches forming the great tuft which terminates the frond, the pinnæ and pinnules are more or less altered from irregular development, the parts being mostly smaller and more generally confluent than in the lower pinnæ, but there is the same setaceous toothing throughout. The extreme points of the upper pinnæ expand into little crispy tufts. It is altogether a most distinct and beautiful variety, represented on our plate, by one of the upper branches, and some of the lower pinnæ detached. The plant was found at Torquay in Devonshire, by Miss Kitson, in 1856. It was communicated to us by Mr. R. J. Gray of Exeter, by whom it is cultivated, and who finds it to be quite constant to its peculiarities. [Plate XXVIII.]

43. *corymbiferum* (M.). This is a corymbosely branched variety of the same general character as *Kitsoniæ* (42), but less marked in its peculiarities. It has been found by Mr. Willison at Whitby, Yorkshire. In this the lower part of the frond is nearly normal with rather small pinnules, the apex being corymbosely branched, but less compoundly so, than in *Kitsoniæ*. It is, however, an elegant variety, and Mr. Willison reports it to be constant.

44. *depauperatum* (Woll.). This rare and curious variety has the fronds frequently so much depauperated that they become mere skeletons with little but the ribs and veins remaining; sometimes a few pinnules are borne in various stages of depauperisation. Occasionally, however, a frond is produced either entirely or partially normal in character. It is, as far as is known, dwarf and barren, though it sometimes produces bulbils. It was found in Ireland—Dublin: Bohernabreena, *Dr. Kinahan*.

GENUS V: LASTREA (*Bory*), *Presl.*

GEN. CHAR.—**Sori** indusiate, globose; the **receptacles** medial or rarely terminal or subterminal on the venules. **Indusium** roundish reniform, sometimes small and irregularly reniform, plane or fornicate, fugacious or persistent, the basal sinus at which it is affixed, variously deep, narrow, broad, or shallow. **Veins** simple forked or pinnate, from a central costa; **venules** free, the anterior usually (sometimes more) fertile.

Fronds herbaceous or coriaceous, pedate pinnate or bi-tri-pinnate, the fertile sometimes slightly contracted.

Caudex short, thick, erect or decumbent; or elongately creeping.

The genus *Lastrea* as established by Presl, embraces that portion of the free-veined species of *Aspidium*, in which the indusium is reniform, and affixed at the basal sinus. It comprises the larger proportion of the free-veined species, and on this ground Professor Fée retains for the group the name of *Aspidium*, which we have already, under *Polystichum*, pointed out, should as it seems to us, rather be preserved to the typical species,—*Aspidium trifoliatum*, a netted-veined plant with umbilicate or peltate indusia, standing in this position. Were it indeed otherwise, the peltate-scaled species have at least the preference over those with reniform covers, so that the *Polystichum* group rather than the *Lastrea* group should retain the original name.

Sir W. J. Hooker adopts a view* different from that of Professor Fée. Retaining the name of *Aspidium* for the peltate-scaled species, without reference to venation we presume, he refers all the species having a reniform indusium to *Nephrodium*, adopting in this respect the views of Richard and Brown. Those who, like ourselves, separate the free-veined from the netted-veined species, assign the

* Hooker, *Filices Exoticæ*, t. 53.

name *Nephrodium* to the net-veined group of reniform *Aspidieæ*; and that of *Lastrea* to the free-veined group.

The name *Lastræa* was first and long ago used by Bory for a sub-generic group, which, neither according to ancient nor modern views can be held to have any value. The name had consequently lapsed, but was revived by Presl under the form of *Lastrea*, for the group now under notice. There is no ground whatever for the arbitrary selection, which Mr. Newman has since made, of *Lastrea Oreopteris* (*montana*), as *the* plant to bear Bory's name, to the exclusion of all the other species now usually associated with it; and this he himself has shown by quoting Bory's subgeneric character, the application of which to this plant was either an original error, or the result of very imperfect observations. *Lastrea montana*, in fact, accords much less exactly with Bory's character, than the three Polypodies he associated with it. Presl was therefore quite justified when in 1836 he revived the lapsed name (altering it to *Lastrea*) for a group which included two of Bory's five species—*Thelypteris* and *Oreopteris* (*montana*),—the others being referrible to *Polypodium*; and though Presl's genus is rather typified by *Lastrea Filix-mas*, it may include without violence the two species just mentioned. There is another name, the *Dryopteris* of Adanson, applied by him to the *Lastrea Filix-mas*, which the botanical-name-reformers of the beginning of the present century would have done well to have adopted; and both Schott in 1834, and subsequently Dr. Asa Gray, have made use of it, but its use has not become extended, and it is therefore hardly binding on us now to revert to such antiquities. Of the two names therefore, applied specially to this generic group, and supported by modern botanical authority, namely *Dryopteris* and *Lastrea*, we select the latter as being the most widely adopted, and therefore avoiding much needless present change; while the old name *Aspidium*, with which Roth's *Polystichum* is nearly equivalent and coeval, is, as we have already mentioned, applied with greater propriety to species having peltate indusia. These several names were judiciously distributed nearly a quarter of a century ago—*Aspidium* to the netted-veined peltate *Aspidieæ*; *Polystichum* to the free-veined peltate *Aspidieæ*; and *Lastrea* to the free-veined reniform *Aspidieæ*; and no further change, at least for the British species, is required.

When the plants now referred to *Lastrea* and *Polystichum* were included in *Aspidium* they bore the English name of Shield Fern. It is however objectionable to use the same appellation for different genera, and as the old name of Shield Fern is properly applied to the more typical group *Polystichum*, we have proposed in the *Handbook of British Ferns* and adopted here, for the *Lastreas*, the equivalent name of Buckler Fern.

The seven or eight British species of *Lastrea* are distributed by Mr. Newman among four genera, of which two originate with himself. Thus, for *Lastrea Thelypteris* he proposes *Hemestheum; Lastrea montana*, better known by the more modern specific name *Oreopteris*, is left to represent *Lastrea*, of which it is assumed to be the type species; while for *Lastrea Filix-mas*, Schott's name *Dryopteris* is adopted; and for *Lastrea cristata* and its allies, *Lophodium* is imposed. These proposed groups do little more than represent specific differences.

The species referred to this genus admit of being distributed into five sectional groups, of which two are represented by British species. These are:—§ *Dryopteris*, and § *Thelypteris*, the former having the veins forked or pinnate, the anterior branch or venule being fertile and the sori medial or subterminal, covered by distinctly reniform indusia, and being represented by *Lastrea Filix-mas* and *Lastrea dilatata;* the latter, having the veins forked, both branches or venules bearing sori, which are covered by irregular fugacious indusia, and represented by *Lastrea Thelypteris* and *Lastrea montana.* The remaining sections are:—§ *Pycnopteris*, with pinnate subclavate veins, and several series of sori, infra-medial on both anterior and posterior venules, represented by the *Lastrea Sieboldii* of Japan; § *Monophlebia*, in which the veins are usually simple, and the sori medial or subterminal, represented by the *Lastrea patens* of South America; and § *Camptodium*, in which the veins are pinnate, and the sori terminal or medial on both anterior and posterior venules, represented by the *Lastrea pedata* of the West Indies.

The name of the genus in its original form (*Lastræa*) was given by Bory de St. Vincent, in honour of M. Delastre of Chatelleraut, a zealous botanist and microscopist. Presl in adopting it and giving it a new application, writes it *Lastrea.*

SYNOPSIS OF THE SPECIES.

Thelypteris.—*Veins usually forked, both branches (anterior and posterior venules) fertile; sori submarginal; indusium irregular, fugacious.*

* *Fronds smooth.*

1. **L. Thelypteris:** fronds not narrowed below, pinnate-pinnatifid; caudex slender, elongately creeping.

** *Fronds resinose-glandular beneath.*

2. **L. montana:** fronds much narrowed below, pinnate-pinnatifid; caudex thick, short, decumbent.

var. **cristata:** fronds corymbosely branched near the apex, the branches as well as the pinnæ multifid-crisped at their apices.

§ **Dryopteris.**—*Veins forked, or sometimes pinnate, the anterior venule fertile; sori medial or subterminal; indusium prominent.*

* *Serratures of pinnules not spinose-mucronate.*

† *Indusium plain or not margined with stalked glands.*

3. **L. Filix-mas:** fronds lanceolate, subbipinnate or bipinnate; pinnules oblong-obtuse, serrate-crenate.

(*a*) *Fronds normal.*

var. **incisa:** pinnules pyramidal-oblong, deeply lobed, the lobes serrate; margin of indusium spreading.

var. **paleacea:** pinnules oblong obtuse, subglaucous beneath, serrate-crenate; rachis densely scaly; margin of indusium incurved.

var. **Pinderi:** fronds narrow elongate lanceolate much narrowed below and above; pinnules, sori, and scales as in *paleacea.*

var. **abbreviata:** dwarf, glandular, subbipinnate; pinnules broad obtuse confluent, crenate-lobate; indusium beaded with glands.

var. **pumila:** dwarf, glandular, subbipinnate; pinnules small obtuse confluent, convex, (pinnæ concave); sori usually uniserial; indusium incurved and beaded with glands.

var. **crispa:** dwarf, crispy, scarcely bipinnate; pinnæ and segments crowded imbricated, wavy, the latter convex, serrate.

var. **Bollandiæ:** fronds ovate-lanceolate, subbipinnate; pinnæ broad; pinnules elongate oblong, obtuse, lobed, somewhat wedge-shaped at the base, and decurrent with the narrowly winged rachis, the lowest only distinct.

(*b*) *Fronds monstrous; apices of fronds and pinnæ multifid-crisped.*

var. **cristata:** normal in size and outline; pinnæ short, narrowed gradually to the much enlarged ramosely-tasselled apex.

var. **polydactyla:** normal in size and outline; pinnæ parallel-sided, abruptly contracted below the small crisp-tasselled apex; pinnules or lobes divided nearly to the rachis throughout.

var. **Jervisii:** normal in size, caudate-acuminate; pinnæ gradually narrowing to the multifid or subcristate apex; upper lobes shallow confluent.

var. **Schofieldii:** fronds dwarf (2-6 in.), frequently ramose or multifid, sometimes crisped, pinnate; pinnæ oblong, lobed or serrate.

†† *Indusium fringed with stalked glands.*

4. **L. rigida**: fronds bipinnate, glandular.

** *Serratures of pinnules spinose-mucronate.*

† *Scales of stipes ovate; indusium without marginal glands.*

5. **L. cristata**: fronds erect, narrow linear, pinnate; pinnæ short triangular; pinnules oblong, all connected, basal ones nearly equal, crenate-serrate or lobed, with aristate teeth.

var. **uliginosa**: fronds erect: the fertile narrow-linear-lanceolate, bipinnate below; pinnules oblong acute, mostly adnate, inciso-serrate or lobed, with aristate teeth, basal ones nearly equal; segments of sterile and autumnal fertile fronds broader.

var. **spinulosa**: fronds erect, narrow-oblong-lanceolate, bipinnate; pinnules oblong acute, posterior basal ones much the longest, all lobed or pinnatifid, with aristate teeth.

†† *Scales of stipes lanceolate, entire or fimbriated; indusium with stalked marginal glands.*

‡ *Scales two-coloured, the centre darker; caudex erect.*

6. **L. dilatata**: fronds lanceolate, ovate or subtriangular ovate, bi-tri-pinnate.

var. **dumetorum**: fronds dwarf, oblong-ovate or triangular, bipinnate, very glandular; pinnules oblong, with coarse teeth; scales broad lanceolate, fimbriate, sometimes pallid.

var. **collina**; fronds narrow ovate elongate, bipinnate; pinnules oblong-obtuse, lobed; lobes obtuse, serrate at end; teeth coarse, acuminate.

var. **Chanteriæ**: fronds lanceolate narrowed below, caudate-elongate at apex; pinnæ distant, the lowest only unequal; pinnules oblong-obtuse, distant, pinnatifid, the lobes coarsely toothed.

var. **angusta**: fronds linear-lanceolate, bipinnate; pinnæ short deltoid, very unequal.

var. **alpina**: fronds narrow linear-lanceolate, membranaceous; scales broader, and paler; sori large; indusium small, fugacious, ragged.

var. **nana**: fronds small, ovate, bipinnate; indusium small, fugacious, slightly glandular.

var. **tanacetifolia**: fronds ample, triangular, tripinnate; indusium small, slightly glandular.

var. **cristata**: fronds and pinnæ dilated, or once or twice irregularly forked or crested at the apex.

‡‡ *Scales pallid whole-coloured, or very indistinctly two-coloured.*

(α) *Caudex erect.*

var. **lepidota**: fronds broadly ovate, quadripinnate; ultimate pinnules small, pinnatifid, spiny-toothed; stipes and rachis everywhere densely scaly.

(β) *Caudex decumbent or creeping.*

var. **glandulosa**: fronds ample, lanceolate-ovate or oblong-lanceolate, tripinnate below, very glandular; scales broad, lanceolate-ovate, semi-appressed.

††† *Scales of stipes lanceolate, crumpled or laciniated; indusium with sessile marginal glands.*

7. **L. æmula**: fronds triangular or triangular-ovate, tripinnate, hay-scented; pinnules concave.

THE FEMALE BUCKLER FERN, or MARSH FERN.

LASTREA THELYPTERIS.

L. fronds lanceolate with a broad base, pinnate, glandless; pinnæ linear-lanceolate, deeply pinnatifid; lobes oblong, their margins revolute in the fertile fronds, the divisions of which thus appear contracted and more acute; caudex creeping. [Plate XXIX.]

LASTREA THELYPTERIS, *Bory, Dict. Class. d'Hist. Nat.* ix. 233. *Presl, Tent. Pterid.* 76, t. 2, fig. 16. *Deakin, Florigr. Brit.* iv. 96, fig. 1605. *Babington, Man. Brit. Bot.* 4 ed. 421. *Newman, Hist. Brit. Ferns,* 2 ed. 183. *Moore, Handb. Brit. Ferns,* 3 ed. 97; *Id., Ferns of Gt. Brit. Nature Printed,* t. 29. *Sowerby, Ferns of Gt. Brit.* 16, t. 7. *Hooker, Gen. Fil.* t. 45 A, fig. 2.

LASTREA PALUSTRIS, *J. Smith, Cat. Cult. Ferns,* 56.

ASPIDIUM THELYPTERIS, *Swartz, Schrad. Journ. Bot.* 1800, ii. 40; *Id., Syn. Fil.* 50. *Smith, Fl. Brit.* 1119; *Id., Eng. Fl.* 2 ed. iv. 272. *Hooker & Arnott, Brit. Fl.* 7 ed. 584. *Mackay, Fl. Hib.* 340. *Bentham, Handb. Brit. Fl.* 629. *Schkuhr, Krypt. Gew.* 51, t. 52. *Willdenow, Sp. Plant.* v. 249. *Sprengel, Syst. Veg.* iv. 104. *Fries, Sum. Veg.* 82. *Fée, Gen. Fil.* 291. *Mettenius, Fil. Hort. Bot. Lips.* 92. *A. Gray, Man. Bot. North U. States,* 2 ed. 597. *Lowe, Nat. Hist. Ferns,* vi. t. 18.

ASPIDIUM PALUSTRE, *Gray, Nat. Arr. Brit. Pl.* ii. 9.

ACROSTICHUM THELYPTERIS, *Linnæus, Sp. Plant.* 1528. *Bolton, Fil. Brit.* 78, tt. 43, 44.

POLYPODIUM THELYPTERIS, *Linnæus, Mant. Plant.* 505. *Hudson, Fl. Ang.* 457. *Flora Danica,* t. 760. *Sturm, Deutschl. Fl.* (*Farnn.*) i. t. 5.

POLYPODIUM PALUSTRE, *Salisbury, Prod.* 403.

POLYPODIUM PTEROIDES, β, *Lamarck, Fl. Franç.* i. 18.

POLYSTICHUM THELYPTERIS, *Roth, Fl. Germ.* iii. 77. *Koch, Synops.* 2 ed. 977. *Ledebour, Fl. Ross.* iv. 513.

NEPHRODIUM THELYPTERIS, *Strempel, Fil. Berol. Synops.* 32.

ATHYRIUM THELYPTERIS, *Sprengel, Anl.* (Eng. ed. 147).

THELYPTERIS PALUSTRIS, *Schott, Gen. Fil.* (sub t. 10, in note.)

HEMESTHEUM THELYPTERIS, *Newman, Phytol.* 1851, *App.* xxii.; *Id., Hist. Brit. Ferns,* 3 ed. 123.

DRYOPTERIS THELYPTERIS, *A. Gray, Man. Bot. North U. States,* 1 ed. 630.

Caudex extensively creeping, sparingly branched, producing fronds at intervals, slender, blackish-brown, scaly at the growing point. *Scales* few, pale brown, ovate-lanceolate. *Fibres* numerous, dark brown, much branched, smooth, or tomentose.

Vernation circinate.

Stipes as long as or longer than the leafy portion in the fertile fronds, less elongated and slighter in the barren; smooth, rounded behind, channelled in front, the base ebony-coloured, pale green upwards; lateral, and adherent to the caudex. *Rachis* also smooth and channelled in front, the secondary rachides bearing a few small scattered scales, and loose spreading deciduous hairs; similar hairs also appear here and there on the veins beneath, and along the margins of the lobes.

Fronds from six or eight inches to four feet in height, including the stipes, and from about four to ten inches in breadth, lanceolate or oblong-lanceolate, scarcely narrowed below, delicate green, membranaceous, erect, pinnate; the barren ones having apparently broader leafy segments, while those of the fertile fronds seem to be narrower and more acute, owing to the rolling in of the margin over or towards the sori. *Pinnæ* numerous, sub-opposite or alternate, spreading, linear-lanceolate, deeply pinnatifid. *Segments* oblong, obtuse or sometimes acute, straight or falcate, entire or slightly sinuate-lobed; the basal ones, especially those on the anterior side, often longer than, and quite distinct from, the rest. The fertile fronds differ from the sterile ones in having the margins of their segments revolute, and in being taller, with stouter stipites.

Venation of the lobes consisting of a stout *costa* or midvein, flexuous in the upper part, from which proceed alternate once or twice forked *veins*, the *venules* running out to the margin. The veins become forked very soon after leaving the midvein.

Fructification on the back of the frond, occupying the whole under surface in the fertile fronds. *Sori* small, round, situated near the base of the venules, *i. e.*, just above the fork of the vein, and forming a line on each side the midvein, and about equally distant from it and the margin, though apparently marginal from the involution of the edge of the frond; they are at first distinct, but often become laterally confluent, and sometimes effused over the whole of the small space between the rolled-up margins. *Indusium* a small delicate roundish-reniform membrane, attached by its posterior edge, the free margin lacerate and glandular. *Spore-cases* numerous, brown, obovate. *Spores* oblong or reniform, strongly muricated.

Duration. The caudex is perennial. The fronds are only of annual duration; the barren ones grow up about May, the fertile in July, all being destroyed by the frosts of autumn.

This plant may be distinguished from the other British species of *Lastrea* by its habit alone, its long, comparatively slender, creeping caudex being unlike that of any other native species; but notwithstanding this, and the fact that its fronds are really quite unlike those of *L. montana* (*Oreopteris*), the species has not unfrequently been confounded with that plant. It will be found to differ from it in having the long creeping caudex just referred to, whilst *L. montana* has a short thick tufted caudex, merely decumbent in habit. It differs further, in having its fronds of their full width almost to the very base, and supported by a long bare stipes, whilst *L. montana* has diminishing pinnæ carried down almost to the base of the stipes; and moreover, in having fronds which are almost free from glands, whilst those of *L. montana* are very conspicuously resinose-glandular on the under surface, and very fragrant. It is still less like any other British species of *Lastrea*.

Lastrea Thelypteris is easily cultivated, merely requiring a light boggy soil, and abundant moisture. Out of doors it should therefore have a damp border, or should be planted in some wettish place about the fernery. A boggy pool at the foot of a mass of rock-work, where it might be accompanied by *Osmunda*, would be a congenial position. In pots, it must have a very abundant supply of water; and the vessels in which it is planted, should be large and shallow, so that its long caudices may have space to spread naturally over the surface of the soil. Peaty soil, alone, or mixed with a proportion of decaying leaves and light sandy loam, will be congenial to it. It is increased readily by division of the caudex.

Though widely dispersed in the United Kingdom, the Marsh Fern is a comparatively rare plant, being local in its occurrence, and growing only in marshy and boggy situations, from some of which it is certainly being displaced by drainage. It is, however, generally abundant where present. In England it is spread from the southern counties to the extreme north, stretching east to Norfolk and west to Somersetshire. In Wales it is found both in the northern and southern divisions. The only Scottish county in which there is certain information of its occurrence is Forfarshire; and the recorded habitats in Ireland are few, though embracing all the provinces of that kingdom. It ranges from the coast level to an elevation of about 600 feet or upwards. The records of its distribution are the following:—

Peninsula.—Devonshire. Somersetshire: Turf Moor, near Bridgewater, very abundant, *T. Clark;* Burtle Moor, *R. Withers.*

Channel.—Hampshire: Portsea; Winchester. Isle of Wight: West Medina; Wilderness; Freshwater Gate; Cridmore, etc. Sussex: Tunbridge Wells; Albourne; Amberley, *W. Borrer;* Waterdown Forest, *W. Pamplin;* Ore, near Hastings.

Thames.—Kent: North Cray; Bexley, *R. Sim;* Ham Ponds, near Sandwich. Surrey: Leith Hill; Hurtmore, near Godalming; Wimbledon Common; Pirbright. Berkshire: Windsor Park and Sunninghill Wells, *J. Bevis.* Essex: Epping; Little Baddow.

Ouse.—Suffolk: Belton; Bungay; Lound; Hipton; Bradwell Common. Norfolk: Horning Marshes; St. Faith's, Newton; Upton Fen; Filby Broad; Holt Lows, *Rev. W. H. Girdlestone;* Edgefield; Scaring Fen; Felthorpe Fen; Wroxham; Dereham; Ormsby Broad; near Yarmouth; Loddon, *Rev. J. J. Smith;* about Norwich, *Rev. W. S. Hore.* Cambridgeshire: Wicken Fen, *Rev. W. H. Girdlestone;* Whittlesea Fen; Teversham Moors; Gamlingay; Fulbourne. Bedfordshire: Potton Marshes. Huntingdonshire.

Severn.—Warwickshire: Bog near Allesley, *Rev. W. T. Bree, W. G. Perry.* Herefordshire. Staffordshire. Shropshire: Whitchurch, *R. W. Rawson;* Berrington Pool, *T. Westcombe.*

Trent.—Nottinghamshire: Oxton Bog; Bulwell Bog. ? Leicestershire.

Mersey.—Cheshire; Knutsford Moor; Newchurch Bog, near Over, *W. Wilson;* Rostherne Moor; Wybunbury Bog; Harnicroft Wood, near Wernith.

Humber.—Yorkshire: Pottery Car, Doncaster; Askham Bog;

Terrington Car; Buttercrambe; Heslington, near York; Settle, *J. Tatham;* Scarborough.

Tyne.—Northumberland: Learmouth Bogs, *Mr. Winch.*

Lakes.—Cumberland: Keswick; Ulleswater; Glencoin; Irton Woods, *J. Robson;* Blowike. Westmoreland: Hammersham Bog.

S. Wales.—Glamorganshire: Singleton Bog; Sketty Bogs, Cwmbola. Pembrokeshire: Pennalle Bog, Tenby, *Rev. W. A. Leighton.*

N. Wales.—Anglesea: Llwydiard Lake, Pentreath, Beaumaris. Carnarvonshire: near Llanberis.

E. Highlands.—Forfarshire: Rescobie; Restenet.

N. Isles.—Shetland.

Ulster.—Antrim: Portmore Park, by Lough Neagh. Galway: Boggy wood at Portumna Castle, *D. Moore.*

Connaught.—Mayo: Near Lough Carra, *J. Ball.*

Leinster.—Wicklow: Marshes at Glencree, *Dr. Mackay.*

Munster.—Kerry: Marshes near Mucrus, Killarney, *Dr. Mackay.*

This Fern extends throughout Europe, being found in the Scandinavian kingdoms, in Russia, in Holland, Belgium, France, Great Britain, Switzerland, Italy, Germany, Hungary, Croatia, Dalmatia. It occurs in North Africa, at Algiers; and a variety, differing only in having a scaly rachis, the *A. squamulosum,* of Kaulfuss, is found at the Cape of Good Hope and Natal, and in New Zealand. The species is recorded as occurring in the Caucasus, and is found among the Altaic ranges of Russian Asia, in Soongaria, and in India in Kashmir. An allied plant of large size, which we regard as only a gigantic variety of this species, has been gathered in Sikkim, by Dr. Hooker. The plant seems to be not unfrequent throughout North America, extending south to New Orleans and Florida; often, however, confused with the sufficiently distinct *Lastrea noveboracensis,* the *Nephrodium thelypteroides* of Michaux.

THE MOUNTAIN BUCKLER FERN.

LÁSTREA MONTANA.

L. fronds lanceolate, much narrowed below, resinose-glandular beneath, pinnate; pinnæ linear-lanceolate, deeply pinnatifid; lobes oblong, flat; sori marginal; caudex tufted; (indusium often obsolete). [Plate XXX.]

LASTREA MONTANA, *Moore, Handb. Brit. Ferns*, 2 ed. 100; 3 ed. 99. *Newman, Hist. Brit. Ferns*, 3 ed. 129.
LASTREA OREOPTERIS, *Bory, Dict. Class. d'Hist. Nat.* ix. 232. *Presl, Tent. Pterid.* 76. *Deakin, Florigr. Brit.* iv. 98, fig. 1606. *Babington, Man. Brit. Bot.* 4 ed. 421. *Sowerby, Ferns of Gt. Brit.* 17, t. 8. *Moore, Handb. Brit. Ferns*, 2 ed. 100; *Id., Ferns of Gt. Brit. Nature Printed*, t. 28. *Newman, Hist. Brit. Ferns*, 2 ed. 188.
ASPIDIUM OREOPTERIS, *Swartz, Schrad. Journ. Bot.* 1800, ii. 35; *Id. Syn. Fil.* 50. *Schkuhr, Krypt. Gew.* 37, tt. 35, 36. *Smith, Fl. Brit.* 1120; *Id., Eng. Fl.* 2 ed. iv. 278. *Bentham, Handb. Brit. Fl.* 629. *Hooker & Arnott, Brit. Fl.* 7 ed. 584. *Mackay, Fl. Hib.* 339. *Willd. Sp. Plant.* v. 247. *Sprengel, Syst. Veg.* iv. 101. *Fries, Sum. Veg.* 82. *Mettenius, Fil. Hort. Bot. Lips.* 92. *Lowe, Nat. Hist. Ferns*, vi. t. 17.
ASPIDIUM ODORIFERUM, *Gray, Nat. Arr. Brit. Pl.* ii. 6.
POLYPODIUM MONTANUM, *Vogler, Dissert. Polypodio montano, cum Icon.* (1787).
POLYPODIUM OREOPTERIS, *Ehrhart, Beitr. zur Naturk.* iv. 44. *Smith, Eng. Bot.* xv. t. 1019. *Flora Danica*, t. 1121.
POLYPODIUM THELYPTERIS, *Hudson, Fl. Ang.* 457. *Bolton, Fil. Brit.* 40, t. 22, fig. 1-2.
POLYPODIUM FRAGRANS, *Hudson, Fl. Ang.* ed. 2, 457 (not of Linnæus).
POLYPODIUM PTEROIDES, *Villars, Hist. des Pl. Dauph.* iii. 841.
POLYPODIUM LIMBOSPERMUM, *Allioni, Auctuar. Fl. Pedem.* 49.
POLYSTICHUM MONTANUM, *Roth, Fl. Germ.* iii. 74.
POLYSTICHUM OREOPTERIS, *De Candolle, Flore Française*, ii. 563. *Koch, Synops.* 2 ed. 978. *Ledebour, Fl. Ross.* iv. 513.
NEPHRODIUM OREOPTERIS, *Desvaux, Ann. Soc. Linn. de Paris*, 257.
PHEGOPTERIS OREOPTERIS, *Fée, Gen. Fil.* 243.
HEMESTHEUM MONTANUM, *Newman, Phytol.* 1851, *App.* xxii.

Var. cristata: fronds and pinnæ multifid-crisped at their apices.

Caudex stout, tufted, decumbent and slowly creeping, formed of the bases of the fronds surrounding a woody axis, scaly. *Scales* pale ferruginous, ovate-acuminate. *Fibres* stout, brown, branching.

Vernation circinate, the pinnæ not folded convolutely.

Stipes short, stout, glandular, and covered with ovate and lanceolate pale brown membranaceous scales; terminal and adherent to the caudex. *Rachis* scaly below, the scales becoming finer and more hair-like upwards; clothed abundantly with sessile glands.

Fronds from one to three feet or more in height, the smaller three inches, the larger eight to ten inches or more in breadth, numerous, erect, terminal, bright green or often yellowish, clothed beneath with a profusion of small sessile resinous glands, which give out an agreeable balsamic fragrance; lanceolate, much tapered towards the base as well as towards the apex, pinnate. *Pinnæ* opposite or alternate, numerous; the lower ones more distant, very short, obtusely triangular; those higher up gradually lengthening till about the middle of the frond, where they are linear-lanceolate, or tapering from a broad base to a long narrow point; the upper ones again are shorter, but also narrower; all deeply pinnatifid. *Lobes* flat, oblong, obtuse, entire or occasionally crenated, sometimes slightly falcate, the basal ones longest.

Venation of the lobes consisting of a flexuous *costa* or midvein, producing alternate *veins*, which are simple or forked; the *venules* extend to the margin, and bear the sori near their apices.

Fructification on the back of the fronds, and most abundant on the upper half. *Sori* moderate-sized, circular, produced near the end of the venules, and forming a submarginal series, often confluent. *Indusium* small, thin, of no definite shape, roundish, jagged, fugacious, often very imperfect, sometimes wanting. *Spore-cases* numerous, brown, obovate. *Spores* roundish or oblong, slightly granulated.

Duration. The caudex is perennial. The fronds are annual, growing up in the spring about May, and becoming destroyed by the autumnal frosts.

This fragrant Fern may be at once distinguished by its balsamic scent; as well as by the short lower pinnæ, which extend down almost to the caudex of the pinnate-pinnatifid, marginally dot-fruited fronds. These fronds grow in tufts. The indusia are generally very small, and soon perish or fall away; they sometimes even appear to be wanting, but the plant is too closely allied to other genuine species of the genus, *L. noveboracensis* for example, to permit of its separation on account of this peculiarity.

We have already mentioned, that there is no good or sufficient reason to fix on this species, as *the* type of Bory's genus *Lastræa;* in fact he altogether ignores the presence of an indusium, which is at least sometimes found in this plant. Bory's genus had lapsed, being an ill-assorted ill-defined group; and when the name in an altered form was revived by Presl, the common Male Fern became its typal representative. The present species is therefore rather an anomalous than a typical member of the group, though not sufficiently so to render its separation necessary.

This species, called the Sweet Mountain Fern, sometimes grows in damp woody places, especially luxuriating by the side of a shady rill or stream; but it is much more profusely met with on the hill sides in heathy mountainous districts. In many parts of the Highlands of Scotland it is the common Fern of the hill sides and road sides. It extends in this direction to the North Highlands and the Western Isles; thence scattered southwards through the Lowlands, it abounds in the Lake districts of the North of England and in Wales, and occurs more or less plentifully in waste districts all over England. It is also found in all the provinces of Ireland; and is, according to Dr. Mackay, plentiful in that country. Mr. Watson gives its range of elevation as extending from the coast level to an altitude of about 2850 feet, which may probably be extended to 3000 feet. The following habitats are recorded:—

Peninsula.—Cornwall. Devonshire: Brendon Wood, and borders of W. Lyn, Lynmouth, *T. Clark;* Barnstaple, *H. F. Dempster;* Challacombe, Exmoor, *H. F. Dempster.* Somersetshire: near Keynsham; Selworthy, *Mrs. A. Thompson,* etc.

Channel.—Hampshire: New Forest near Lyndhurst; near Southampton. Isle of Wight: Apse Castle. Dorsetshire. Wiltshire. Sussex: Danny, near Brighton, *Rev. T. Rooper;* Tilgate Forest, *J. A. Brewer;* Waterdown Forest; Eridge Woods, near Tunbridge Wells, and elsewhere.

Thames.—Hertfordshire: Bell Wood, Bayford; Tring; Broxbourne, etc. Middlesex: Hampstead. Kent: Bexley, *R. Sim;* Blackheath; Bailey's Hill, between Brasted and Tunbridge, *W. Pamplin;* Tunbridge Wells. Surrey: Witley; Houndsdown Bottom, near Hindhead; Cobham; Wimbledon, etc. Oxfordshire: Shotover Hill. Buckinghamshire: Hartwell. Essex: High Beech; Little Baddow, *A. Wallis.*

Ouse.—Suffolk: Bradwell. Norfolk: near Crome, *R. Wigham.* Cambridgeshire: Fulbourne, Teversham, etc. Northamptonshire: Dallington Heath.

Severn.—Warwickshire: Allesley; about Arbury Hall; Coleshill Heath; Corley, *Rev. W. Bree;* Dunsmore Heath, near Rugby, *Ray.* Gloucestershire: Ankerbury Hill, Forest of Dean, *W. H. Purchas;* Leigh Woods, near Bristol. Monmouthshire: Glyn Ponds; Nantygollen, near Pont-y-pool, *T. H. Thomas.* Herefordshire. Worcestershire: Malvern Hills, *E. Lees;* Bromsgrove. Staffordshire: Ramshaw Rocks, near Warslow. Shropshire: Whitcliffe; Ludlow; Shawbury Heath; Wyre Forest.

Trent.—Leicestershire: near Twycross. Rutland. Lincolnshire. Nottinghamshire: Hartswell, near Farnsfield; Oxton Bog; Eddingley Bog. Derbyshire: Dethich Moor; Riley, *Dr. Howitt.*

Mersey.—Cheshire: Moorland, near Birkenhead, *F. Brent;* Oxton. Lancashire: near Warrington, *W. Wilson;* Middleton, *H. Buckley;* Rochdale; Rainhill; Gateacre.

Humber.—Yorkshire: Sheffield; Valley of the Don, near Doncaster; Melton Wood, near Adwick; Escrick, near York; Whitby; Castle Howard Woods; Richmond; Bradford, *J. T. Newboult;* Halifax; Everley, near Scarborough, *W. Bean.*

Tyne.—Durham: Chapel Weardale; Darlington; Cawsey Dene, near Newcastle; by the Tees. Northumberland: Embleton; banks of the Irthing, *Rev. R. Taylor.*

Lakes.—Cumberland: Keswick; near Lodore waterfall, *H. C. Watson;* Patterdale; Hawl Gill, Wastwater. Westmoreland: Rydal Water; Langdale, and other parts. Isle of Man.

S. Wales.—Radnorshire. Brecknockshire. Glamorganshire: Swansea, *T. B. Flower.* Carmarthenshire. Cardiganshire.

N. Wales. — Anglesea. Denbighshire: Wrexham; Llanymynech, *C. C. Babington.* Flintshire. Merionethshire: Dolgelly.

Carnarvonshire: near Llanberis; Nant Gwynedd, *C. C. Babington;* Aber, and other parts.

W. Lowlands.—Dumfriesshire: Moffat Dale, *P. Gray.* Lanarkshire.

E. Lowlands.—Roxburghshire: Ruberslaw. Edinburghshire: Pentland Hills; Habbie's How, *T. M.* Berwickshire: Dye at Longformacus; Banks of Whiteadder.

E. Highlands.—Stirlingshire: Ben Lomond. Clackmannanshire: Dollar, *J. T. Syme.* Kinross-shire. Fifeshire: Lomond Hills, *Dr. Balfour.* Forfarshire: Glen Isla; Clova Mountains; Sidlaw Hills. Perthshire: Pass of Trosachs, *T. M.;* Dunkeld; Craig Chailliach; Ben Lawers; banks of Loch Tay; and elsewhere, abundant. Aberdeenshire, *Dr. Murray.* Morayshire.

W. Highlands.—Argyleshire: Glencroe; Glen Gilp, Ardrishiag; Ballenoch, *T. M.;* and elsewhere, common. Dumbartonshire: Tarbet, plentiful. Bute: Rothesay, *T. M.* Isles of Arran, Islay, and Cantyre.

N. Highlands.—Sutherlandshire, common, *Dr. Johnston.*

N. Isles.—Shetland, *Cyb. Brit.*

W. Isles.—N. Uist.

Ulster.—Donegal: Milroy Bay, *E. Newman;* Killybegs, *R. Barrington;* near Lough Eske, *R. Barrington.* Londonderry, *D. Moore.*

Connaught.—Galway: Lough Corrib; Connemara; between Dooghty and Maam; ascent of Maam Turc Pass; Letterfrank, *E. T. Bennett.*

Leinster.—Wicklow: Glencree, *S. Foot;* Seven Churches, *D. Moore;* Glendalough, and Powerscourt.

Munster.—Clare: between Innistymon and Corrafin, *E. T. Bennett;* Feacle, *J. R. Kinahan.* Waterford: near Clonmel. Kerry: Mangerton; Killarney, *S. P. Woodward.*

This species is met with throughout the whole of Europe, being recorded from Norway, Denmark, Holland, Belgium, France, Spain, Italy, Switzerland, Germany, Hungary, Croatia, Transylvania, Greece, and extending eastwards to Moscow. We have a memorandum of its occurrence at Pico, one of the Azorean Islands being probably intended; and there exists in the Hookerian herbarium a specimen labelled from North America, and one from Vermont in that of the Rev. W. A. Leighton. In Chili and Valparaiso a closely allied plant, differing only in being slightly hairy, is found.

Though so common a species in some localities, it is not one which readily submits to cultivation, and many have been the failures of those who have attempted its domestication. Mr. Wollaston has suggested a mode of treatment which is quite in accordance with its natural habits. The plan is to pot or plant in pure loam, and to keep this soil wet through the winter, this being done, if the plants are potted, by keeping a feeder full of water constantly beneath them. Probably a continuous supply from a syphon, allowing the superfluous quantity to overflow, so that there might be a constant change going on, would be a still better arrangement; it would at least assimilate more exactly with the ceaseless percolation which must be going on its native hills. We have succeeded tolerably well by following this plan; though in the smoky climate of London, the plants seldom retain their vigour for any length of time. In country situations, where the atmosphere is pure, the plants grow tolerably well if the situation is rather moist; and under such circumstances, they are better exposed than confined. There is no difficulty in securing a supply of the plants in the localities where the species occurs, young seedling plants being generally most abundant. The caudices also grow in tufts, which may be separated for propagation.

The Sweet Mountain Fern is not very prolific of varieties. There are, however, a few varied and curious as well as distinct forms, which have the greater interest from being found to retain their peculiarities. These are:—

1. *truncata* (Woll.). This is a curious monstrosity, and proves quite permanent under cultivation. The apices of the fronds, and with very few exceptions, those of the pinnæ also, terminate abruptly, and the end of the rib or rachis projects, often nearly a quarter of an inch, beyond the pinnules. The plant has thus the appearance of having had the ends of its pinnæ and its apex, eaten off in a uniform manner by some mollusk, and the graceful outline and

aspect proper to the species in its normal state are quite destroyed. In all other respects this variety resembles the normal plant. It has been found—near Tunbridge Wells, Kent, *G. B. Wollaston*. Carnarvonshire: Llanberis, *Rev. J. M. Chanter*. [Plate XXXI.]

2. *caudata* (M.). This has a good deal of the normal character, but it differs conspicuously from the common states of the plant, in having the apex of its fronds and of the pinnæ narrowed and drawn out to a considerable length, so that the fronds become elongately lanceolate, and the pinnæ caudate. It was found at Windermere by Mr. Clowes, and has been cultivated by him without changing its character for several years. The apex of the frond is frequently, but not constantly, divided in a trifid manner.

3. *abrupta* (M.). This form, the constancy of which has not been sufficiently tested, has narrow fronds; the pinnæ are short, bluntish, often bifidly and very obtusely dilated at the end, the pinnules interrupted and irregular in size. It was found by Mr. C. Jackson, near Barnstaple.

4. *crispa* (M.). This form has the pinnules undulated or wavy, so that the frond acquires a crispy appearance; it is otherwise normal. It was found by Dr. Balfour, on the Clova mountains.

5. *cristata* (M.). This is a beautiful variety, analogous to the cristate forms of *Filix-mas* and *Filix-fœmina*, and most nearly resembling *Athyrium Filix-fœmina, var. corymbiferum* in its conformation. The fronds are rather small, and corymbosely branched towards their apex, the branches being dilated into broad cristate tassels; the apices of the pinnæ also are multifid-crisped, though in a less degree than the apex of the frond itself. It was found in Monmouthshire, by Mr. T. H. Thomas.

THE MALE FERN, or COMMON BUCKLER FERN.

LASTREA FILIX-MAS.

L. fronds lanceolate, subbipinnate or bipinnate; pinnules oblong, obtuse or acutish, serrate crenate or inciso-lobate, the basal ones more or less distinct, the upper confluent; serratures not spinulose; indusium convex, persistent, (and except in *abbreviata* and *pumila*) without marginal glands.

—(type) subbipinnate; pinnules oblong, obtuse, with a broad attachment or connected at the base, crenate-serrate (chiefly at the apex), green beneath; sori usually extending from the base about half the length of the pinnules. [Plate XXXII.]

Lastrea Filix-mas, *Presl, Tent. Pterid.* 76. *Deakin, Florigr. Brit.* iv. 103, fig. 1609. *Babington, Man. Brit. Bot.* 4 ed. 421. *Sowerby, Ferns of Gt. Brit.* 19, t. 9. *Moore, Handb. Brit. Ferns,* 3 ed. 102; *Id., Ferns of Gt. Brit. Nature Printed,* t. 14. *Newman, Hist. Brit. Ferns,* 2 ed. 197.

Polypodium Filix-mas, *Linnæus, Sp. Plant.* 1551. *Bolton, Fil. Brit.* 44, t. 24. *Hudson, Fl. Ang.* 458.

Polypodium nemorale, *Salisbury, Prod.* 403.

Aspidium Filix-mas, *Swartz, Schrad. Journ. Bot.* 1800, ii. 38; *Id., Syn. Fil.* 55. *Smith, Fl. Brit.* 1121; *Id., Eng. Bot.* xxi. t. 1458; and xxviii. t. 1949 (excl. text); *Id., Eng. Fl.* 2 ed. iv. 275. *Schkuhr, Krypt. Gew.* 45, t. 44. *Hooker & Arnott, Brit. Fl.* 7 ed. 585. *Mackay, Fl. Hib.* 340. *Bentham, Handb. Brit. Fl.* 629. *Fée, Gen. Fil.* 291. *Flora Danica,* t. 1346. *Svensk Bot.* t. 51. *Fries, Sum. Veg.* 82. *Hooker, Fl. Lond.* iv. t. 40. *Willd. Sp. Plant.* v. 259. *Sprengel, Syst. Veg.* iv. 105. *Mettenius, Fil. Hort. Bot. Lips.* 92, t. 18, fig. 7. *Lowe, Nat. Hist. Ferns.* vi. t. 13.

Aspidium Blackwellianum, *Tenore, Att. Accad. del. R. Inst. Sc. Nat. Nap.* v. (reprint 9, t. 3, fig. 9.)

Aspidium nemorale, *Gray, Nat. Arr. Brit. Pl.* 7.

Polystichum Filix-mas, *Roth, Fl. Germ.* iii. 82. *De Candolle, Fl. Franç.* ii. 559. *Koch, Synops.* 2 ed. 978. *Ledebour, Fl. Ross.* iv. 514.

Nephrodium Filix-mas, *Richard, Cat. Med. Paris,* 1801, 120. *Hooker, Fil. Exot.* under t. 98. *Lowe (R. T.), Trans. Camb. Phil. Soc.* vi. 527.

Tectaria Filix-mas, *Cavanilles, Prælect.* 1801, 251.

Dryopteris Filix-mas, *Schott, Gen. Fil.* (sub. t. 9). *Newman, Hist. Brit. Ferns,* 3 ed. 183, fig. *b.*

Lophodium Filix-mas, *Newman, Phytol.* 1851, *App.* xx.

Var. **incisa**: fronds robust, bipinnate; pinnules elongate or pyramidate-oblong, acutish, deeply inciso-lobate, the lobes serrate; sori usually occupying nearly the whole pinnule. [Plate XXXIII B.]

Lastrea Filix-mas, *v.* incisa, *Moore, Phytol.* iii. (1848) 137; *Id., Handb. Brit. Ferns,* 3 ed. 103; *Id., Ferns of Gt. Brit. Nature Printed,* t. 15. *Babington, Man. Brit. Bot.* 3 ed. 410.

LASTREA FILIX-MAS, β. AFFINIS, *Babington, Man. Brit. Bot.* 4 ed. 421.
LASTREA EROSA, *Deakin, Florigr. Brit.* iv. 101, fig. 1608 (excl. syn. *Aspidium erosum*, Schkuhr; according to the figure t. 45).
LASTREA AFFINIS, *Moore MS.* (Nat. Pr. Ferns).
ASPIDIUM FILIX-MAS, β. EROSUM, *Hooker & Arnott, Brit. Fl.* 6 ed. 569; 7 ed. 585 (excl. *Aspidium erosum*, Schkuhr). *Döll, Rhein. Fl.* 16.
ASPIDIUM DEPASTUM, *Schkuhr, Krypt. Gew.* t. 51 (monstrous).
ASPIDIUM AFFINE, *Fischer & Meyer, Hohen. Enum. Talüsch.* 10. *Fée, Gen. Fil.* 291. *Ruprecht, Dist. Crypt. Ross.* 36.
ASPIDIUM CAUCASICUM, *A. Braun, Flora*, 1841, 707.
ASPIDIUM PSEUDO-FILIX-MAS, *Fée, Iconogr. Nouv.* 103.
ASPIDIUM FILIX-MAS, *Hohenacker, Enum. Elizabethpol.* 260. *Tenore, Att. Accad. del R. Inst. Sc. Nat. Nap.* v. (reprint 7, t. 1, fig. 1 A-B.)
ASPIDIUM MILDEANUM, *Goeppert, Bot. Zeit.* xii. 85.
POLYPODIUM HELEOPTERIS, *Borkhausen, Röm. Arch. Bot.* i. 19; according to Deakin.
POLYSTICHUM AFFINE, *Ledebour, Fl. Ross.* iv. 515.
POLYSTICHUM FILIX-MAS, *var.* 1, *Roth, Fl. Germ.* iii. 84.
LOPHODIUM EROSUM, *Newman, Phytol.* 1851, *App.* xxi.
DRYOPTERIS AFFINIS, *Newman, Hist. Brit. Ferns*, 3 ed. 187 (183, fig. *a*).
DRYOPTERIS FILIX-MAS, *v.* AFFINIS, *Newman, Hist. Brit. Ferns*, 187.

Var. paleacea: fronds subbipinnate; pinnules oblong, truncately obtuse, serrate at the apex, paler subglaucous and hair-scaly beneath; sori distinct, often small, confined to the lower part of the pinnules; margin of the indusium much inflected beneath the spore-cases; stipes and rachis shaggy with lustrous golden-brown scales, long and narrow above, intermixed with broader ones at the base. [Plate XXXIII A.]

LASTREA FILIX-MAS, *v.* PALEACEA, *Moore, Handb. Brit. Ferns*, 2 ed. 110; 3 ed. 103; *Id., Ferns of Gt. Brit. Nature Printed*, t. 17 A.
LASTREA FILIX-MAS, *v.* BORRERI, *Johnson, Sowerby Ferns of Gt. Brit.* 20.
LASTREA PSEUDO-MAS, *Wollaston, Phytol.* n. s. i. 172.
LASTREA PALEACEA, *Moore MS.* (Nat. Pr. Ferns).
LASTREA PATENTISSIMA, *Presl, Tent. Pterid.* 76.
LASTREA PARALLELOGRAMMA, *Liebmann, Mex. Bregn.* 119.
LASTREA TRUNCATA, *Brackenridge, United St. Explor. Exped.* xvi. 195, t. 27.
ASPIDIUM PALEACEUM, *Don, Prod. Fl. Nepal.* 4.
ASPIDIUM PATENTISSIMUM, *Wallich, Cat.* 340 (scales darker).
ASPIDIUM DONIANUM, *Sprengel, Syst. Veg.* iv., *part* 2, 320.
ASPIDIUM WALLICHIANUM, *Sprengel, Syst. Veg.* iv. 104.
ASPIDIUM PARALLELOGRAMMUM, *Kunze, Linnæa*, xiii. 146.
ASPIDIUM CRINITUM, *Martens & Galeotti, Foug. Mex.* 66, t. 17, fig. 2. *Fée, Gen.* 292.
ASPIDIUM AFFINE, *A. Braun, Flora*, 1841, 707, in obs.
ASPIDIUM NIDUS, *Griffith MS. in Hb. Hook.*
ASPIDIUM ULIGINOSUM, *Blume MS. in Hb. Hook.*
ASPIDIUM ADNATUM, *Blume, Enum. Fil. Jav.* 162.
NEPHRODIUM AFFINE, *Lowe, (R.T.) Bot. Misc.* n. s., i. 25; *Id., Trans. Camb. Phil. Soc.* vi. 525.

NEPHRODIUM PATENTISSIMUM, *Strachey & Winterbottom, Hb. Himal.* 5.
NEPHRODIUM FILIX-MAS, *v.* PALEACEUM, *Hooker, Fil. Exot.* t. 98.
DICHASIUM PATENTISSIMUM, *A. Braun, Flora,* 1841, 710. *Fée, Gen. Fil.* 303, t. 23 B, fig. 2 (stipes).
DICHASIUM PARALLELOGRAMMUM, *A. Braun, Flora,* 1841, 710. *Fée, Gen. Fil.* 303, t. 23 B, fig. 1.
DRYOPTERIS BORRERI, *Newman, Hist. Brit. Ferns,* 3 ed. 189.
DRYOPTERIS FILIX-MAS, *v.* BORRERI, *Newman, Hist. Brit. Ferns,* 3 ed. 189.

Var. **Pinderi**: fronds narrow elongate lanceolate, much attenuated at both the base and apex, subbipinnate; pinnules, scales of stipes, and sori as in *paleacea,* the basal scales elongate subulate.

LASTREA FILIX-MAS, *v.* PINDERI, *Moore, Pop. Hist. Brit. Ferns,* 2 ed. 315; *Id., Handb. Brit. Ferns,* 3 ed. 103.

Var. **abbreviata**: fronds dwarfish, glandular, subbipinnate, the pinnæ concave, scarcely pinnate; pinnules, large (comparatively), broad, obtuse, mostly decurrent, unequally crenate or crenate-lobate, the lobes with blunt teeth; sori usually uniserial on each side the midrib; indusium fringed with glands; scales somewhat fim briated.

LASTREA FILIX-MAS, γ. ABBREVIATA, *Babington, Man. Brit. Bot.* 3 ed. 410; 4 ed. 421. *Johnson, Sowerby Ferns of Gt. Brit.* 20. *Moore, Ferns of Gt. Brit. Nature Printed,* under t. 14; *Id., Handb. Brit. Ferns,* 3 ed. 103.
LASTREA ABBREVIATA, *Moore MS.* (Nat. Pr. Ferns).
POLYSTICHUM ABBREVIATUM, *De Candolle, Fl. Franç.* 3 ed. ii. 560.
ASPIDIUM ABBREVIATUM, *Poiret, Encyc. Bot. Supp.* iv. 516.
LOPHODIUM ABBREVIATUM, *Newman, Phytol.* 1851, *App.* xxi.
DRYOPTERIS ABBREVIATA, *Newman, Hist. Brit. Ferns,* 3 ed. 192.
DRYOPTERIS FILIX-MAS, *v.* ABBREVIATA, *Newman, Hist. Brit. Ferns,* 3 ed. 192.

Var. **pumila**: fronds dwarf, glandular, subbipinnate; pinnæ deflexed, concave; pinnules small, convex, mostly confluent, bluntly crenate-serrate; sori usually confined to the *lowest* anterior venule of the *lowest* pinnules, and thus arranged in a single series on each side the midrib of the pinnæ; indusium somewhat inflected at the margin, and beaded with short-stalked glands; scales slightly fimbriated. [Plate XXXV.]

LASTREA FILIX-MAS, *v.* PUMILA, *Moore, Ferns of Gt. Brit. Nature Printed,* t. 17 B; *Id., Handb. Brit. Ferns,* 3 ed. 104.
LASTREA PUMILA, *Moore MS.* (Nat. Pr. Ferns).
LASTREA ABBREVIATA, *Wollaston, Phytol,* n. s., i. 172.
LASTREA FILIX-MAS, *v.* ABBREVIATA, *Moore, Handb. Brit. Ferns,* 2 ed. 103.
ASPIDIUM PUMILUM, *Lowe, Nat. Hist. Ferns,* vi. t. 15.
ASPIDIUM FILIX-MAS, *v.* RECURVUM, *Francis, Anal. Brit. Ferns,* 38.
ASPIDIUM FILIX-MAS, *v.* PUMILUM, *of gardens.*

Var. **Bollandiæ**: fronds ovate-lanceolate, minutely glandular, subbipinnate; pinnæ broad oblong, acutish; pinnules elongate oblong, obtuse, deeply lobed, the base narrowed but decurrent with the wing of the rachis, the basal ones only being distinct; scales fimbriate.

Var. **crispa**: fronds dwarf, crispy, scarcely bipinnate; pinnæ crowded, overlapping, wavy, the points curving upwards; segments imbricated, convex, serrate.

LASTREA FILIX-MAS, *v.* CRISPA, *Sim, Cat. Ferns*, 1859, 10.

Var. **cristata**: fronds and pinnæ symmetrically multifid-crisped at the apex; pinnæ narrowed gradually towards, and much constricted near the large ramosely multifid-crisped tassel; pinnules blunt, subglaucous, as in *paleacea*. [Plate XXXVI.]

LASTREA FILIX-MAS, *v.* CRISTATA, *Moore & Houlst., Gard. Mag. Bot.* iii. 317. *Moore, Handb. Brit. Ferns*, 2 ed. 106; 3 ed. 104; *Id., Ferns of Gt. Brit. Nature Printed*, t. 16 A.

Var. **polydactyla**: fronds and pinnæ multifid-crisped at the apex; pinnæ much and suddenly narrowed just behind the crispy tassel; pinnules or lobes divided nearly to the rachis throughout, linear-oblong, dilated or acute, serrated. [Plate XXXVII.]

LASTREA FILIX-MAS, *v.* POLYDACTYLA, *Moore, Ferns of Gt. Brit. Nature Printed*, t. 16 B; *Id., Handb. Brit. Ferns*, 3 ed. 104.

Var. **Clowesii**: fronds and pinnæ multifid-crisped at the apex; pinnæ narrowed gradually to the tufted apex, pinnate below, pinnatifid above with deep bluntish lobes.

Var. **Jervisii**: fronds caudate-acuminate and as well as the pinnæ multifid and subcristate at the apex; pinnæ narrowed gradually, not suddenly contracted; pinnules or lobes oblong acutish, serrate, those towards the apices of the pinnæ shallow, so that the portion of the pinna below the tuft is not much divided, but acutely serrate.

LASTREA FILIX-MAS, *v.* JERVISII, *Moore, Handb. Brit. Ferns*, 3 ed. 112.

Var. **Schofieldii**: fronds dwarf (2-6 inches), frequently ramose or multifid, sometimes crisped, pinnate; pinnæ oblong, lobed or serrated. [Plate XXXVIII.]

LASTREA FILIX-MAS, *v.* SCHOFIELDII, *Sim, Cat. Ferns*, 1859, 10.
LASTREA DILATATA, *v.* SCHOFIELDII, *Stansfield MS. Moore, Ferns of Gt. Brit. Nature Printed*, under t. 22; *Id., Handb. Brit. Ferns*, 3 ed. 138.
LASTREA SPINULOSA, *v.* SCHOFIELDII, *Sim, Cat. Ferns*, 1856, 5.

Caudex large, tufted, scaly, erect or decumbent, often in age becoming considerably elongated, consisting of the bases of the old fronds persistent around a woody axis, from the apex of which appear the growing fronds. *Scales* like those at the base of the stipites. *Fibres* protruding from among the bases of the fronds at the lower part of the caudex, strong, coarse, dark-coloured, deeply penetrating.

Vernation circinate, the apex of the frond becoming liberated before the whole rachis is uncurled, and at this stage bent downwards with a curve resembling that of a shepherd's crook.

Stipes short, stout, from three to six inches long, terete, slightly channelled in front, densely clothed with large scales of a narrow lanceolate attenuate outline, membranaceous chaffy texture, and pale brown colour, intermixed with smaller and shorter ones; terminal, and adherent to the caudex. *Rachis* clothed sparingly with small subulate scales.

Fronds averaging two or three feet in height, but varying from a foot to four or six feet, according to age, variety, and locality; and when the crown is vertical, arranged in a circlet around it; erectish, herbaceous, smooth, of a lively rather deep green, somewhat paler beneath; broadly lanceolate with a gradually tapering apex, or sometimes oblong lanceolate with a sudden acumination at the apex; bipinnate. *Pinnæ* numerous, alternate or nearly opposite, linear, gradually narrowing towards the apex, which is acute; the lower ones decreasing in length from about the middle of the frond, the lowermost measuring an inch or rather more in length in fronds of a foot and a half long, those about the middle being three or four inches long; the lower pinnæ are also more distant than those higher up. *Pinnules* at the base of the pinnæ distinct, notched on both sides at the base but with a broad attachment, or sometimes slightly connected, the first pair somewhat larger; the rest generally attached by the entire width of their base, and more or less combined, with a very narrow sinus; oblong obtuse, *i. e.*, of equal width throughout, with the apex rounded, slightly crenate or crenate-lobed at the margin, serrated around the blunt apex, the teeth acute but not spinulose.

Venation of the pinnules consisting of a flexuous *costa* or midvein bearing alternate branches or *veins*, which are again branched once or twice, these secondary branches or *venules* extending nearly to the margin, each venule (or vein) itself if simple, or the anterior branch if ramified, proceeding towards the point of one of the marginal serratures, just within which it terminates. The manner of ramifying is, by what is called forking, which consists in the production of two branches both slightly and about equally diverging from the straight line. In the larger varieties there are more of these forkings than in the smaller.

Fructification on the back of the frond, rarely extending more than half-way down, and most copious on the upper third. *Sori* numerous distinct, roundish-reniform, in the normal form confined to the lower half of the pinnules, attached to the anterior venule at a short distance above its source, and much below its termination, thus becoming medial on the vein, and forming two short lines extending upwards from the base of the pinnæ, rather nearer the midvein than the margin. *Indusium* firm, convex, persistent, reniform, *i. e.*; roundish with a posterior notch, affixed by the notch or sinus, with an entire margin, *i. e.*, without marginal glands, (except in *pumila*, and *abbreviata*, which are probably distinct) and acquiring a grayish or leaden hue as the fructification becomes matured. *Spore-cases* reddish brown, obovate. *Spores* oblong or reniform, granulated.

Duration. The caudex is perennial. The young annual fronds are produced about May, and endure throughout the summer and autumn, and until destroyed by severe frost. Under shelter all the forms are subevergreen, *paleacea* and its subvarieties most especially.

This plant is the type of the modern genus *Lastrea*, which consists of indusiate free-veined dot-fruited Ferns, having the indusium reniform, *i. e.*, round with a notch in the margin, forming a sinus, by which it is affixed.

The common Male Fern cannot well be mistaken for any other native species. It has indeed been formerly confounded with *L. cristata*, but the two have no very close affinity, and the only near resemblance occurs in a form of *Filix-mas*, not common, in which the lower pinnæ are triangular. The Incised variety is in some respects like *L. rigida*, but obviously different in many others.

This is one of our most common and most widely-dispersed Ferns, growing abundantly in sylvestral and rupestral situations over the whole of England, Wales, Scotland, and Ireland, as well as in the Northern and Western Isles, and in Guernsey and Jersey. According to Mr. Watson, it ascends to an elevation of 1500 feet in the Highlands, but is rare above the agrarian zone. The varieties *incisa* and *paleacea* have been gathered in so many, and such widely separate localities, that there is reason to believe them nearly, if not quite as generally dispersed as that we have taken as the typical plant. These three forms are indeed so common that, notwithstanding their very obvious differences, it is probable that many persons take them indifferently for the common *Filix-mas*.

The Male Fern appears to be abundant over the whole of Europe; as are also, probably, the Incised and the Golden-scaled forms of it. Thus, for example, the range of the species is known to extend in Europe from Scandinavia and Russia to Spain, Italy, and Crete; and in Asia from the Ural Mountains and the Caucasus, to the Siberian chain of the Altai and the neighbourhood of Lake Baikal, occurring about Erzeroum in Armenia, and extending along the chain of the Himalaya, from Kumaon through Nepal to Assam. It is also found in Northern Africa, and in Madeira; and, in the New World, in Newfoundland, California, Mexico, Guatemala, New Grenada, Equador, Peru, Brazil, and the Caraccas. It is not found, we believe, in the United States. It is difficult to allot exactly their several stations to the three principal forms assumed by this common plant, but it is most probable that each of them has a very similar range. The Incised form, *Aspidium affine* of Fischer and Meyer, is found in Russia in various places in the region of the Caucasus and in Georgia—Elizabethpol, Karabagh, Lenkoran; and in the New World, in Mexico. The Golden-scaled form, varying with darker, often very dark-coloured scales, represented by the names *Aspidium paleaceum* of Don, and *A. patentissimum* of Wallich, is found in several parts of India—the Himalaya, Kumaon, and Mussoree; in Ceylon, and in Java; in Madeira, where it seems to be the prevailing form; and in various parts of America (often with the scales very dark rather than golden-brown), as in Mexico and Guatemala, in Columbia, Peru, Quito, and Brazil.

The Male Fern has long had, and still retains, a reputation as a medicine, its use being as an anthelmintic. Theophrastus and Dioscorides, by whom it was called *pteris*, and Pliny, who calls it *Filix-mas*, as well as Galen, all appear to have used it as such.* The attention of modern practitioners became directed to it principally from the circumstance of its being one of the remedies employed against tape-worm by Madame Nouffer, who sold the secret of her method of treatment to Louis XVI. for 18,000 francs. The 'fern root' had, however, apparently fallen into disuse, at least in this country, probably from the substitution of other sorts for the true plant, or in consequence of other more efficient agents, especially oil of turpentine, having been found; but from a recent account by Dr. Lindsay,† it would appear again to have come into use. Dr. Lindsay's remarks are to this effect:—

"This [*Lastrea Filix-mas*] has been repeatedly used, of late, in different wards of our hospital as an anthelmintic in the treatment of tape-worm (*Tænia solium*). It has also been extensively applied to the same purpose by the profession in Edinburgh and other parts of Scotland. It had fallen into disuse greatly in this neighbourhood in consequence of supposed inefficiency, but undeservedly so, until Professor Christison, in two papers, 'On the Treatment of Tape-worm by the Male Shield Fern,' published in the *Edinburgh Monthly Medical Journal* (June, 1852; July, 1853), showed that want of success, in some cases, depended on bad preparations of the root, or on old roots, being used. He found it almost uniformly successful in the form of an oleo-resinous extract, obtained by percolation of the root with ether. It is recommended in the dose of eighteen to twenty-four grains, followed by a purgative. In many parts of England nothing is more common as a vermifuge than half a drachm to a drachm of the powder of the root made

* Pereira, *Elements of Materia Medica*, 577.

† Dr. W. Lauder Lindsay, in *Phytologist*, iv. 1062.

up in the form of an electuary with a little treacle or jelly. In other parts of the country the oil of the Male Fern is an equally common nostrum. But in neither of the latter conditions can its action be relied on, especially if purchased in the shops of druggists, who generally not only sell old roots and bad preparations, but some the roots of totally different species. It is most apt to be, and has most frequently been confounded with *Athyrium Filix-fœmina*, the root of which it has yet to be proved has a similar virtue. This fern was first used at Geneva by Peschier, some twenty or thirty years ago, in the form of an etherial extract, but it appears to have been recommended as a vermifuge by Theophrastus, Dioscorides, and Galen; and it formed the chief part of Madame Nouffer's celebrated remedy for the tape-worm. It does not appear to be accurately determined on what special ingredients of the root its vermifuge property depends. We know it contains tannic and gallic acids. There is some contrariety of opinion as to the proper period of the year for collecting the plant for use; Peschier regarding it as most effectual if gathered between May and September, and Professor Christison considering the date of collection immaterial. The only caution necessary in using it is probably that it ought always to be had fresh; if gathered and prepared by the practitioner himself so much the better. The oleo-resin, however, seems to retain its properties for a considerable time; though what this period accurately is still remains *sub judice*. It has been found quite efficient after being kept a year. Professor Christison commends it as a less disagreeable and more efficient anthelmintic than the 'Abyssinian Kousso, the Continental Pomegranate, or the American Turpentine.' It is surprising that Peschier's observations, made on a very large scale indeed, have attracted so little attention in Britain."

It is the caudex which is the part to be employed. Dr. Pereira describes this correctly as being almost completely enveloped by the thickened bases of the footstalks of the fallen leaves; and the fern root of the shops, he says, consists of fragments of the dried thickened bases of footstalks to which small portions of the stem are found adhering. The caudex and footstalks are in the recent state fleshy and of a light yellowish-green internally, but in the dried state they are reddish-white. When fresh, iodine colours them bluish-black, indicating the presence of starch, particles of which substance may be recognised by the microscope. The dried 'root' has a feeble earthy somewhat disagreeable odour. Its taste is at first sweetish, then bitter, astringent, and subsequently nauseous like rancid fat. The caudices should be collected in the height of summer, but both in the whole state and powdered they deteriorate by keeping. Large doses of it would appear to excite nausea and vomiting. The anthelmintic property of the drug resides in the oil, which forms but a small proportion of the bulk, as will be seen by the following analysis by Geiger,* quoted by Dr. Pereira:—

Ligneous fibre and starch	56·3
Incrystallisable sugar / Oxidisable tannin	22·9
Gum and salts, with sugar and tannin	9·8
Green fat oil	6·9
Green resin	4·1—100·0.

According to another analysis of Morin, † quoted in the same work, the constituents of the Fern stems are:—volatile oil; fixed oil (stearin and olein); tannin; gallic and acetic acids; incrystallisable sugar; starch; gelatinous matter insoluble in water and alcohol; ligneous fibre; and ashes, consisting of carbonate sulphate and hydrochlorate of potash, carbonate and phosphate of lime, alumina, silica, and oxide of iron.

The Male Fern is also applied to various economic uses, such as the bleaching of linen, the manufacture of glass, and the tanning of leather. The Bracken and the Male Fern were in Lightfoot's time burnt together for the sake of their ashes, which were used by the soap and glass makers; and he mentions that in the Island of Jura, 150*l.* worth of these ashes was exported annually. The astringent stems are used in dressing leather, and the ashes in bleaching linen. Bishop Gunner relates,‡ that the young curled fronds on their first appearance out of the ground are boiled and eaten like asparagus, and that the poorer Norwegians cut off the succulent laminæ

* Geiger, *Handb. de Pharm.* 1829.
† Morin, *Journ. de Pharm.* x. 223.
‡ *Flora Norvegica.*

at the crown of the root, which are the bases of the future stalks and brew them into beer, adding thereto a third portion of malt. In times of great scarcity they mix the same with their bread. He adds, that if cut green and dried in the open air, this Fern affords not only an excellent litter for cattle, but that if infused in hot water, it becomes no contemptible fodder for goats, sheep and other cattle, which readily eat, and sometimes grow fat upon it,—a circumstance, Lightfoot observes, well worth the attention of the inhabitants of the Highlands and the Hebrides, as great numbers of their cattle, in hard winters, frequently perish for want of food. The dried fronds are, moreover, a good protective material for plants.

The stems of this Fern, with their young incurved and yet unexpanded fronds, have been turned to superstitious use, the St. John's hands, or 'lucky hands,' being prepared from them, and sold to credulous and ignorant folk as preservatives against witchcraft and enchantment. These preparations are figured under the name of 'Johannis Hand,' by Schkuhr, at t. 46 *b*, fig. 1, of his work on Ferns.

The culture of the Male Fern is not at all difficult. It may be grown in any cool shady place, in almost any kind of soil, the best being a sandy loam, moist, but not wet. It may be planted with good effect about shady walks, in woods and wilderness scenery, and on shady rockwork. The variety *incisa* is very striking, where effect only, and not variety, is the object; and fine plants of the variety *paleacea* are very noble in appearance. Potted plants require ample space, and should be plunged out-doors in winter. Though a common, it is a very handsome plant, and one of the most desirable in large ferneries. It is increased by division.

It may be mentioned as a curious fact, that the permanent so-called varieties of our hardy Ferns are very generally, if not in every case, reproduced from their spores, and in most cases abundantly so. The crested form of *Filix-mas*, the multifid *Asplenium Trichomanes*, and some of the most remarkable of the forms of *Scolopendrium*, have been raised by hundreds in this way. The fact of reproduction from the spores has been in some instances considered as the test of a species, and it is a test to which one would, at first thought, be inclined to submit; but the experience of

Fern-growers shows it to have no value whatever among Ferns. *Cystopteris Dickieana*, which has the aspect of a species, is reproduced from its spores, and this fact might seem to prove its distinctness; but the forms referred to above are clearly not species, but varieties of well-known species, and they, too, are reproduced with equal constancy. So that the test of reproduction from the spores, fails as the mark of a species. The fact itself is probably suggestive that spores are bodies rather of the nature of buds than of seeds.

The Male Fern occurs commonly in three distinct forms, as has been already mentioned, these forms being probably often taken indifferently as representatives of the species. There are besides them, moreover, a considerable number of forms which are recognised as varieties of a secondary character, many of which are permanent, and objects of much interest. In addition to several cristate varieties, all of which are very handsome, there are some striking forms of the *incisa* type, a singular elongated form of the *paleacea* type, and a beautiful dwarf crisped variety, which perhaps belongs to the *pumila* type.

1. *erosa* (Clowes). This form has oblong obtuse biserrated pinnules, which are more or less and variously abbreviated or erose in the fertile portions, giving an irregularly contracted appearance to these parts; while the lower sterile portions are sometimes much like *rigida*. The fronds are, however, at other times more lax and more generally affected, the pinnæ being then distant, and the pinnules more widely separated and decurrent, as well as more irregular in size. When in this state, it has a semi-depauperated aspect. It has been found at Lodore, near Keswick, by Miss Wright, and at Windermere by Mr. Clowes, who finds it subpermanent.

2. *interrupta* (M.). This is a curious abnormal-looking form, very irregular in development. The pinnules are in great measure changed from the normal character, sometimes forming short inciso-serrate lobes along the rachis, sometimes larger and laciniate, the fronds having a tendency to divide at the apex. The general effect is that most of the pinnules are very much narrowed. It was found at Windermere by Mr. Clowes, and is a constant form.

3. *biformis* (M.). This a dwarf form, accidentally raised from spores in the Chelsea Botanic Garden. It produces two forms of fronds, a portion being dwarf and normal, and the rest much depauperated, the pinnules being very small and confluent, so that the pinnæ are linear, with distant minute marginal lobes, the larger of which are serrated. It has remained for three years constant to these peculiarities.

4. *incisa* (M.). This is altogether a larger and more striking plant than the normal form, of which it is probably the full development. It is of robust, stately habit, averaging three or four feet, but sometimes reaching six feet in height, with a stipes of five or six inches in length. The fronds in unfolding liberate the point, which becomes bent like the curve of a shepherd's crook, as in the common plant; they are distinctly bipinnate, lanceolate, not contracting abruptly near the apex. The pinnæ are elongate, tapering gradually to the apex. The pinnules are somewhat less closely placed; the basal ones notched, often deeply, on each side their base, thus having a narrow attachment, elongately pyramidate-oblong, broadest at the base, and with a narrow though rounded apex; the rest more broadly attached, and more equal in width; the margins more or less deeply inciso-lobate, the lobes three to five-toothed. The venation is more highly developed than in the common form, thus:—a vein is directed up the centre of each lobe, and this bears alternately several venules; but the sori are, notwithstanding, produced only on the anterior basal venule of each fascicle, so that, as in the normal form, they are ranged in a single line on each side the midvein, commonly extending, however, much nearer to the apex of the pinnule. The indusium is here reniform as in the other, convex, entire, and persistent, but not inflected as in *paleacea*. The irregularly deformed monstrous leafy developments of this variety constitute the *Aspidium depastum* of Schkuhr. [Plate XXIII B.—Folio ed. t. XV.]

This variety is probably equally common with the type form, and appears as widely dispersed. The following habitats are known:—

Peninsula.—Devonshire: Lindridge, Teignmouth, *Miss A. Hooseason;* Combe Martin, *C. C. Babington.* Somersetshire: Bridgewater; Nettlecombe, *C. Elworthy.*

Channel.—Sussex: Hurst Green, *Rev. J. Hand.* Dorsetshire: Bridport. Wiltshire.

Thames.—Hertfordshire: Barnet. Kent: Tunbridge, *M. T. Masters;* Sturry. Surrey: Reigate; St. Martha's Hill, Guildford, *T.M.;* Mayford; Woking;. Merrow; Moor Park, near Farnham, *Mrs. Walker.*

Ouse.—Bedfordshire: Sutton, and Potton, *R. Heward.* Norfolk: Norwich, *E. Field.* Northamptonshire: King's Cliffe.

Severn.—Herefordshire: Copped Wood Hill, near Ross, *W. H. Purchas.* Worcestershire: Malvern; Daylesford Hill, *H. Buckley.*

Trent.—Nottinghamshire: Wollaton. Derbyshire: Stapenhill.

Tyne.—Cumberland: Cockermouth. Northumberland: Bedlington, near Morpeth, *Rev. R. Taylor.*

E. Lowlands.—Berwickshire: The Dene, Bogan Green, Coldingham, *R. Hogg.*

W. Lowlands.—Lanarkshire: Cathcart Hill, near Glasgow, *Dr. Deakin.*

E. Highlands.—Kinross-shire. Perthshire: Dunkeld; Ben Chonzie, near Crieff. Kincardineshire: Kingcausie, *J. T. Syme.*

W. Highlands.—Argyleshire: Cairndow.

Leinster.—Dublin: Kingstown, *R. Barrington.*

Channel Isles.—Guernsey, *C. Jackson.*

A monstrously developed state, in which the parts are broader and more decurrent, and frequently irregular in development, but which is probably only an accidental development of this variety, has been gathered at Teignmouth, Tunbridge, Stapenhill, Dolgelly, Durham; Balliivy, Down; and in Guernsey.

Another form (*deorso-lobata*), in which the pinnules have an enlarged posterior basal lobe, is probably a peculiar form of *incisa;* the pinnules are oblong and obtuse, the basal ones with a narrow attachment; the margins of the basal ones are more or less inciso-lobate, with the lobes serrated, and the lower posterior one much enlarged and forming a kind of auricle directed towards the main rachis, which is the chief peculiarity of the variety, and occurs in various lesser degrees in most of the common states of this type. The rest of the pinnules are more or less inciso-serrate. This is the variety *spinosum* of the earlier editions of Mr. Francis's *Analysis of British Ferns*, but is not at all spinose. It appears to be common, and to be in fact one of the two larger forms into which the species is commonly developed, the characteristics of the two being sometimes united in one plant. The Rev. W. A. Leighton appears to have first noticed the peculiar lobing; his specimens gathered twenty years since are from Bomere and Sutton in Shropshire, and from Anglesea. We have also seen specimens from—Penryn, Cornwall; Cobham Park, and Maidstone, Kent; Mayford, Sutton Park, St. Martha's Hill, and Bagshot, Surrey; Epping, Essex; Lynn, Norfolk; Black Park, Bucks.; Salisbury, Wilts.; Daylesford, Worcestershire; Matlock, Derbyshire; Leeming Lane, Bedale, Yorkshire; Ambleside, Westmoreland; Castle Malgwyn, Pembrokeshire; Ruthin, Denbighshire; Callender and Kinnoul Hill, Perthshire; Glen Gilp and Cairndow, Argyleshire; Arran; Tinnahinch, Wicklow; Ballyvaughan, Clare; Killarney, Kerry; Athenry, Galway; and from Jersey and Guernsey.

5. *producta* (M.). This is a very striking variety, somewhat analogous to *incisa*, being like it a large growing plant, with fronds at least three feet long and ten inches broad, and also resembling it in the divided condition of the pinnules, which are however much more deeply divided. The frond is lanceolate; the lowest pair of pinnæ two inches and a half long, triangular acuminate; the next are more than four inches long, triangular elongate, being nearly two inches across the base, gradually tapering to a sharp elongated point; those in the upper part of the frond, though narrower than those below, are also triangular elongate, the base being broadest, though in the uppermost the sides become more nearly parallel. The pinnules throughout are elongated, deeply pinnatifid, and narrow upwards to the apex, their outline being that of a narrow cone or pyramid; this, together with the deep and conspicuous lobing, gives to the plant a very distinct appearance. The basal pinnules, which are cut down more than half way to the midrib, have an attachment so narrow as to resemble a mere winged petiole; the rest, half way up the pinnæ, though adnate, are attached by less than their whole width. The lobes of the pinnules are obscurely serrated at the end. The sori are confined to the

upper third of the frond; and on the fronds we have seen, have a manifest tendency to occupy rather the central than the basal portion of the pinnules: that is, they are distant from the base on those pinnules which occupy the lower half of the fertile pinnæ. This variety has been found—in Shropshire: on the Wrekin, *Rev. W. A. Leighton.* Buckinghamshire: Black Park, *Dr. Allchin.* Surrey: Portnall Park, Virginia Water, *T. M.* Kent: Pitt's Wood, Chislehurst, *G. B. Wollaston;* Varnes, *Dr. Allchin.* Devonshire: Barnstaple, *C. Jackson.* It may be considered as a finely developed state of the *incisa* type, with elongated pyramidal pinnules. [Plate XXXIV B.]

6. *elongata* (M.). This is a large growing and beautiful form, of the *incisa* type. The fronds are fragrant, lance-shaped, with rather distant caudate pinnæ. The pinnules are elongate or linear-oblong, narrow as compared with their length, somewhat falcate, serrate on the margin, the apex subacute, and they possess the peculiarity of being irregular in length, which gives an unevenness of outline to the pinnæ not usual in forms that are not depauperated. The scales of the stipes are fringed. It was found in the Isle of Wight by the Rev. W. H. Hawker, and Mr. A. G. More; and the same form has been gathered at Addington, Gloucestershire, by Mr. H. Buckley. [Plate XXXIV A.]

7. *latipes* (M.). A very curious form of thick fleshy texture, and having two kinds of fronds. The most normal looking is about two feet high, ovate, the lower pinnæ not abbreviated, coarse, scarcely bipinnate; the oblong overlapping almost spiny-serrate segments or pinnules being connected by a narrow wing. The more abnormal frond is larger, with large distant deeply inciso-lobate much decurrent pinnules or segments, which are very irregular in size and form, and are in some parts depauperated. The upper part of the frond is fertile. We have no information as to its constancy; but it is a very marked variation. It was found at Ruthin in Denbighshire, by Mr. T. Pritchard; and a similar plant but smaller, also dimorphous, has been found in Guernsey by Mr. C. Jackson.

8. *triangularis* (M.). This form belongs to the *incisa* group, but has something of the aspect of *Lastrea cristata,* being remarkable for its narrow stiff erectish fronds, and especially for the unequally

triangular outline of a few of the lower pinnæ, the lowest pair especially, these having more of the outline met with in *L. spinulosa* than that which usually occurs in *L. Filix-mas*. The pinnules are longish, and all but the basal ones are adnate or decurrent; they are inciso-serrate or lobed, and they sometimes show the enlarged posterior basal lobe, which occurs in *deorso-lobata*. It has been found in several parts of Kent, from whence we are indebted for specimens to Dr. Allchin. The most marked is from Holt Wood, near Maidstone, Kent.

9. *paleacea* (M.). This variety, which is a widely dispersed, and probably not uncommon form, is subbipinnate, the basal pinnules only being distinct; the pinnules are oblong, somewhat glaucous beneath, truncately-obtuse, and serrated at the apex. The fronds are broad-lanceolate, one to five feet high, having the scales of the stipes broad-lanceolate, while those of the rachis, and of the pinnules are hair-like. The pinnæ are pinnate. The pinnules oblong, obtuse, serrated at the apex, with a broad attachment. One of its most obvious characteristics is the subglaucous under surface, the other forms of the species being green. It is often of a yellowish green colour, but not always, being frequently deep green, but the under surface is always paler. When in its most fully developed state it differs from the normal form in the abundance of the lustrous golden-tinted scales, which clothe its stipites and rachides, so densely that their rich colouring is conspicuous, especially at the back of the frond. The scales are however not always so richly-coloured nor always abundant, being sometimes dark-coloured, and scattered, but they always differ obviously from the pallid scales of the other common forms. Mr. Wollaston points out, that the rachides and midveins are more or less tinged with purple, but this also occurs sometimes in *incisa*. The sori are usually, if not constantly, smaller, and the indusium before maturity, and even when the spore-cases are ripening, has its margins very much inflected beneath them, so that, when reversed, it is seen to have the form of a little pouch, as in the leaflets of *Cheilanthes lendigera*. In the common and incised forms of *Filix-mas*, on the contrary, the margin of the indusium is merely bent down a little sloping outwards till it comes in contact with the surface of the pinnule. [Plate XXXIII A.—Folio ed. t. XVII A.]

This plant appears to be not uncommon, and is widely dispersed both in the East and in the New World. The known British habitats are as follows :—

Peninsula.—Cornwall : Penryn, abundant, *G. Dawson* ; Devonshire : Torquay ; Barnstaple, *C. Jackson.*

Channel.—Sussex : Hastings, and also in other parts, *Dr. Allchin* ; Uckfield, *N. B. Ward.* Wiltshire : Salisbury.

Thames.—Kent : Tunbridge Wells, *G. B. Wollaston,* and elsewhere. Surrey : Albury ; Chobham.

Severn.—Worcestershire. Shropshire : Linley, near Broseley, *G. Maw.*

Humber.—Yorkshire.

Tyne.—Durham.

Lakes.—Westmoreland : Ambleside, *Miss Beever.*

S. Wales.—Cardiganshire : Hafod, *E. Newman.* Pembrokeshire : Castle Malgwyn, *W. Hutchison.*

N. Wales.—Denbighshire : Ruthin, *T. Pritchard.* Montgomeryshire, *Mrs. Walker.* Merionethshire : Dolgelly, at the base of Cader Idris, *T. Hankey.*

E. Highlands.—Stirlingshire : Polmont ; Inversnaid. Perthshire : Pass of the Trosachs, *T. M.* ; Ben Lawers, *T. M.* Forfarshire : Clova Mountains, *J. Backhouse.*

W. Highlands.—Dumbartonshire : Tarbet. Argyleshire : Glen Gilp, Ardrishiag ; Loch Ballenoch ; Glen Croe, *T. M.* ; Cairndow ; Glen Kinglass. Isle of Arran.

Ulster.—Donegal : Killybegs, *R. Barrington.*

Connaught.—Sligo : Lough Gill, *R. Barrington.* Galway : Kylemore, *R. Barrington.* Mayo : Ballycroy Mountains, *R. Barrington.*

Leinster.—Wicklow : Glencullen, and near Upper Lough Bray, *R. Barrington* ; near the Dingle, *Dr. Allchin.*

Munster.—Kerry : Cahir Conree, near Tralee, *Dr. Allchin.*

Channel Isles.—Guernsey, *C. Jackson.* Jersey.

It is, in part, from the Indian forms of this plant, called *Aspidium patentissimum* by Dr. Wallich, and differing from our European form in nothing except the darker colour of their scales, and their somewhat larger growth, that Prof. Braun has constituted his genus *Dichasium,* which is characterised by having 'biscutelloid' indusia, which are indusia of roundish outline with a sinus extending upwards beyond the centre, so that the lobes look like saddle flaps. We have ascertained from a careful examination of Dr. Wallich's specimens that this appearance of the indusia is merely the result of age. In the younger and perfect state the indusium is round, convex, with a posterior notch or sinus, and very much inflected margins, just as occurs in the British plant. As the spore-cases enlarge, they are unable to lift off the indusium, in consequence of its constricted margin, and the result is that the edge becomes split opposite the sinus. The indusium is then pushed up by the advancing spore-cases, the upper margin is apparently brought nearer the point of attachment, and the two halves assume the appearance which attracted the attention of Prof. Braun. Exactly the same structure of indusium occurs in the other species referred to this supposed genus—a Columbian plant collected by Hartweg.

There are of this as of the var. *incisa,* numerous examples of monstrous development, which in this variety, commonly takes the form of a branching of the rachis, a forking of the apex, or a branching of the points of the pinnæ. Such growths are not usually permanent. The var. *multifida* of which the peculiarities are that the apex of the frond, and of more or fewer of the pinnæ are bifid or multifid, and that the pinnæ are at the same time occasionally depauperated to a mere rib, is one of these monstrous forms, which is nearly permanent under cultivation. We have examples of these abnormal growths from Penryn, Cornwall; Barnstaple, Torquay, and Upcott, Devon; Tunbridge Wells, Kent; Settle and Castle Howard Woods, Yorkshire; Dublin county; and from Guernsey.

We have received from Mr. Jackson a broad leafy form, gathered at Barnstaple, in which the pinnules, without being much changed in form, are irregularly and unequally laciniate-lobate. This, which belongs to *paleacea,* may if constant, be called *fissum.*

10. *paleaceo-lobata* (M.). This is a very remarkable and beautiful form of *paleacea,* in which the margins of the pinnules are lobate, and also somewhat undulated. The fronds are very large, and the pale or glaucous under-surface and rich coloured scales are well marked. The finest, because undulated form, was found at Tarbet,

Dumbartonshire; and we have other lobate forms from Penryn, Cornwall; Glen Lochy, Perthshire; Ardrishiag, Argyleshire; and from Jersey and Guernsey. [Plate XXXIII C.]

11. *paleaceo-crispa* (M.). A handsome subvariety of the *paleacea* type. The fronds of this form are broad or ovate, densely-leafy, the pinnæ and pinnules, the latter especially, being remarkably close set. The chief peculiarity, however, resides in the undulation of the pinnules, these being twisted and curled so as to give quite a crispy appearance to the surface of the frond; they are rather elongate-oblong in outline, the basal ones lobate, and the tips of all of them sharply serrated. Though a plant of full size, even the lowest pinnules are not quite distinct, a narrow wing to the rachis connecting the rest with each other; in the case of this lowest pinnule, the sinus is continued so as to narrow the attachment, but the upper pinnules are attached nearly or quite by their whole width. It was found by Mr. R. Hogg, at Bogan Green, Coldingham, Berwickshire.

12. *Pinderi* (M.). This is a remarkable and elegant form, peculiar from its long narrow fronds, which are very much attenuated both towards the base and apex, thus becoming elongate-lance-shaped in outline. The fronds are nearly or quite a yard in height, and less than six inches wide in the broadest part, tapering upwards into a long slender point, and narrowed below in a similar way. The stipes is short. This belongs to the golden-scaled type, the pinnules scales and sori being similar to those of that variety, the chief difference consisting in the remarkable outline of the frond. The lowermost scales are, however, very long and subulate. It was found near Elter Water in the Lake district, in 1855, by the Rev. G. Pinder.

13. *abbreviata* (Bab.). This dwarf-growing form, seldom exceeding a foot in height, is one of the permanently smaller forms of the species, and is probably specifically distinct; though *pumila* (14) has many characters in common with it, and the two are perhaps forms of one subalpine species. The present is however a larger plant than *pumila*, with considerably larger, broader, and therefore coarser-looking pinnules, which although to some extent recurved, are yet by no means so fully or so constantly so, as in *pumila*. The scales of the stipes are somewhat fimbriated or jagged at the margin.

The fronds, at least while young, are glandular and fragrant; they are pinnate; the pinnæ scarcely again pinnate, the lowest pinnules only being sometimes separated, the remainder always decurrent; the points of the pinnules are turned upwards so that the upper surface of the pinnæ is concave. The pinnules are large for the size of the plant, broad, rounded at the apex, the margin unequally crenate, or crenate-lobate, the lobes having blunt obscure teeth. It is allied to the variety *pumila*, but differs in the larger size of its pinnules, which gives it a coarser aspect, and it is not so much recurved. The sori are for the most part uniserial on each side the midrib of the pinnæ; and have indusia which at least while fresh, are margined with glands, as in *pumila*. This rare form has been found in the habitats below named—Snowdon: Cwm Glas, *Rev. J. M. Chanter;* also in the same district, *W. Pamplin*. Durham: Teesdale, *J. Backhouse*. Yorkshire: Ingleborough, *Rev. G. Pinder*. Lancashire: Conistone, *Miss Beever*. Westmoreland, *G. B. Wollaston*. Cumberland, *Rev. G. Pinder*. Gloucestershire: Wyck, *Bab. Man.* Forfarshire: Glen Isla, *J. Backhouse*. Kerry: Killarney, *R. Barrington*.

14. *pumila* (M.). This is a permanently small dwarf erect plant, remarkable among other characteristics, for the recurving of the points of its pinnæ, and of its pinnules, which gives to its upper surface a concave appearance. It usually grows from nine inches to a foot in height, and rarely, when very vigorous, reaches the height of a foot and half; the stipes being two to three inches long, and furnished with scales which are fimbriate on the margin. The fronds are lanceolate, pinnate. The pinnæ are short, bluntish, rather deflexed, scarcely ever more than deeply pinnatifid, the basal pinnules only being sometimes but rarely semi-detached. The pinnules or lobes are small oblong, obtuse, obscurely crenated, convex, but recurved at the points, so that the pinnæ are concave, the points of the pinnæ being also recurved, so that the frond itself is concave. The venation is comparatively simple: the costa or midvein, which is carried up each lobe, produces veins of which the lower are once forked, the upper simple. In fronds of ordinary growth, scarcely any but the anterior branch of the lowest anterior vein in each lobe or pinnule bears a sorus, and the sori then form an almost simple

line on each side the midrib of the pinnæ about even with the sinuses of the pinnules. When however the growth is very luxuriant, a few of the basal pinnules bear two, three, or four sori each, but even in these cases, the sori form two simple series for more than half the length of the pinnæ, so that the general uniserial arrangements is hardly disturbed. The indusium is convex, reniform, and persistent, and its margin is somewhat inflected beneath the spore-cases; it is, moreover, beaded with short-stalked glands. We are almost persuaded that this plant offers specific differences, in its constantly small size, the direction of the pinnæ and pinnules, the peculiar distribution of the sori, the glandular inflected indusium, and in the important character of vernation. In respect to this latter point, the shepherd's crook form, which occurs in the common Male Fern, is not assumed in the process of unrolling its fronds, but the rachis gradually unrols from the base to the apex. It is also reproduced from the spores, although that alone is not evidence of its distinctness. On the other hand, the general character of the parts, and of the sori and indusia, agree with diminutive examples of the Male Fern. The fresh fronds are fragrant, in consequence no doubt of the presence of numerous small glands on their surface; the fragrance having something of the sweetness of Mignonette. This rare Fern appears to have been first brought from Snowdon, by the late Mr. D. Cameron, and has been recently found near Llyn Ogwen by Mr. S. O. Gray. Mr. Wollaston finds it sometimes dichotomously divided at the apex. It seems confined to North Wales, and to alpine localities. [Plate XXXV.—Folio ed. t. XVII B.]

15. *crispa* (Sim). This a beautiful dwarf evergreen Fern, remarkable for its crispy surface. We have not seen it in a mature fertile state, but the plants appear to resemble *pumila* (14) in size and habit, and they agree with that variety in having fimbriated scales. The fronds are lanceolate, with an acuminate apex. The pinnæ, which are thickly set upon the rachis, so that they overlie each other, are deeply pinnatifid, but scarcely pinnate even at the base. The segments are oblong, crowded, overlapping, and rather distinctly serrated. The stipes is short, and very scaly, as is the main rachis behind; smaller scales are also scattered along the secondary

rachides. The main rachis is entirely hidden in front by the crowded imbricated pinnules. The chief peculiarity, after its dwarfness and density, and that to which it mainly owes its distinctness and beauty, is the undulation of the parts, which gives it a fine crispy appearance. This is produced by the points of the pinnæ turning upwards or backwards from the plane of the rachis, so that the surface of the frond is concave, while the convex segments are unequally deflexed, producing an uneven convexity of the surface of the pinnæ. It is one of those forms which have proved themselves constant from the spores, a large number of plants having been raised. We are indebted to Mr. R. Sim for our knowlege of the plant, and the annexed particulars of its history:—"It was gathered in Wales, by Mr. J. W. Salter of London, and by him given to the Hon. Mrs. Wrightson of Warmsworth Hall, Doncaster, in whose fernery the original plant still exists. It was there seen by Mr. S. Appleby of Balby, who struck with the peculiar habit, obtained a fertile frond, from which plants have been raised, all exactly resembling the parent." We have lately received from Miss Stancomb of Trowbridge, a similar form, found in 1858 in the neighbourhood of South Molton, Devonshire, by the Rev. T. Mann; this also proves constant, and it agrees with the Welsh plant in everything but the distinctly serrated margins.

16. *Bollandiæ*. (M.). This very interesting variety has, when fresh, a remarkably powerful mignonette-like fragrance, resembling that of *pumila* and *abbreviata*, arising doubtless from numerous minute sessile or embedded glands, the fronds appearing to be glandular-punctate. The fronds grow about a foot and a half in height, and are ovate-lanceolate, membranaceous, and bipinnate at the very base of the pinnæ, but there only, all the other pinnules being connected by the wings of the rachides. The pinnæ are broad, oblong, stalked, narrowed rather suddenly at the point, not at all acuminate. The pinnules are large, elongate-oblong, obtuse, pointing forwards, somewhat wavy, deeply lobed, the lobes sparingly serrate, the base narrowing in a wedge-shaped manner, and becoming decurrent with the narrow but distinct membrane which borders the secondary rachides. The scales of the stipes are very pale brown, and distinctly fimbriated. It was found in a hedge-row near

Ashurst Park, Tunbridge Wells, in 1857, by Mrs. Bolland, and is cultivated in the collections of Mr. Hankey and Mr. E. A. De Grave, of Fetcham, Surrey. The breadth of the fronds, the winged rachides, and the large wavy pinnules, give it a peculiarly distinct and interesting character.

17. *cristata* (M.). This is doubtless the most beautiful, all points considered, among the British Ferns; for its tall, gracefully arching, symmetrically and boldly tasselled deep-green enduring fronds are certainly unsurpassed in elegance, and this notwithstanding that it is in a botanical sense, a monstrosity. Like many others of the monsters, however, that occur among the Ferns, it is reproduced almost without variation from the spores. The fronds are narrow lanceolate, with short rather distant pinnæ, which are narrowish and taper from the base upwards to the base of the tuft or tassel. This tassel, which occurs at the end of every pinna, consists of a large branched tuft of multifid-crisped segments, forming a very conspicuous frilled margin to the frond. The apex of the frond is also branched and multifid-crisped like the pinnæ, only the tuft is larger and more distinctly ramose. The symmetrical character of this frilling, is one of its most beautiful features. The pinnules are oblong obtuse, subglaucous beneath, the stipites and rachides being golden-scaly, and the indusia inflected at the margin, as in *paleacea* (9), to which this form is evidently allied. The typical state of this variety, and the most beautiful so far as yet known, was found at Charleston, near St. Austell, in Cornwall; but other similar plants have been found in Devonshire, near Ilfracombe, by Mr. J. Dodds. Young plants are commonly symmetrical, though sometimes irregularly ramose, but they all at length assume the same characteristic form. [Plate XXXVI—Folio ed. t. XVI A.]

A sub-variety of this (*prolifera*, Woll.) raised from its spores, resembles it in being tasselled, but it is more crisped, and is depauperated and laciniated. Its chief peculiarity consists in its bearing bulbils, generally on the external side of the stipites, near their junction with the tufted caudex.

18. *cristata angustata* (M.). This is a handsome new variety, raised from the spores of *cristata*, and now sufficiently tested as to its constancy, the parent plant having been grown for several years,

and young ones quite characteristic abundantly produced from it. It differs from the older *cristata* (17), very much in the same way as *Pinderi* differs from *paleacea*, namely in the elongation and coincident narrowing of the fronds. It is however still more remarkable, for it is only pinnate, the pinnules being all confluent, so that the margins of the pinnæ are only shallowly lobed. The largest fronds we have seen, which we believe are of mature size, are about eighteen inches long and two and a quarter inches broad. The pinnæ are consequently so short, that the frond seems to consist only of a frill on each side of the rachis. The pinnæ have an enlarged semicordate base, representing the basal pinnules; above this they are contracted, with a lobate-serrate margin, and they terminate in a roundish crispy tuft; the elongated apex of the frond also terminates in a multifid-crispy tuft of considerable size. This elegant variety was raised by Mr. R. Sim, of Footscray, Kent.

19. *dentex* (M.). A curious variety raised from *cristata* by the Messrs. Stansfield, of Todmorden. It has the apex of the frond constantly truncate or abrupt, the usual attenuated upper portion being replaced by an ordinary sized pinna; and the apices of the pinnæ are variously bifid or ramose, but not crispy. The frond is bipinnate; the pinnules oblong, close-set, and deeply and sharply serrated, the toothing being conspicuous.

20. *Jervisii* (M.). This is a tasselled state allied to the typical form of the species. Like that it has a broadly lanceolate frond, with an acuminated apex. The pinnæ are elongated and rather irregular at the margin, distinctly divided below into oblong serrated acutish pinnules, but towards the apex merely cut into shallow acute lobes, which point forwards and extend to the base of the tassel, which is dilated and subcristate, but less tufted than in the other cristate forms. The plant is very elegant, and has a peculiar and distinct aspect, resulting from the mode in which the upper part of the pinna is lobed. The plants, while young, are often not very characteristic, but they become so as they get older. This is the *polydactyla* of most collections of living plants, the parent having been at first identified with that variety, from which it proves to be materially different. It was found by Mr. Swynfen Jervis, in the vicinity of his residence, Darlaston, near Stone, Staffordshire.

21. *Clowesii* (M.). This is another tasselled form, referrible to the type form of the species, but quite distinct from *Jervisii* (20), in the more crispy tassels, and the more deeply pinnatifid attenuated pinnæ. The fronds are lanceolate, not suddenly acuminate. The pinnæ are elongated, and gradually narrowing for their whole length up to the base of the tassel; pinnate in the basal portion, and deeply pinnatifid throughout the remaining part, with blunt-ended lobes, the lobes being tolerably distinct and obtuse quite up to the tassel. The tassel itself forms tolerable sized multifid-crispy tufts; and the apex of the frond is ramosely multifid-crisped. It differs from *polydactyla* (22) in having less elongated pinnules, and tapered pinnæ, the tassel being about equally developed. This variety was found at Windermere, by Mr. F. Clowes.

22. *polydactyla* (M.). This is a tasselled form referrible to the *incisa* group. The pinnæ are not shortened as in *cristata*, nor are they tapered gradually towards the end as in *Jervisii* and *Clowesii;* in fact they do not narrow much until quite close to the tassel which terminates each of them. The pinnules are incised, and the basal ones have a tendency to dilatation at their tips. The apex of the frond is more or less tufted. Here and there a frond has its apex more decidedly tasselled with a corymbose ramification, and the pinnæ themselves are more normal, merely indicating a tendency to division; while occasionally a frond is produced in which the tendency to laceration both in the pinnæ and pinnules is carried to excess, becoming grotesque. It was found at Bromsgrove, in Worcestershire, and was communicated by B. Maund, Esq. [Plate XXXVII.—Folio ed. t. XVI B.]

23. *furcans* (M.). The fronds of this variety are of normal character, except in this, that the ends of the pinnæ are forked, usually once but sometimes twice, the divisions being short, tapering, and divergent, so as to produce a curious fish-tail-like appearance. It was found in the neighbourhood of Huddersfield, Yorkshire, by Mr. T. Stansfield, and a similar form has been gathered on Bookham Common, Surrey, by Mr. Stedman.

24. *Schofieldii* (Sim). This is a curious pigmy plant, which has been a puzzle to many experienced pteridologists, by whom it has been successively referred to *L. dilatata, spinulosa,* and *Filix-mas.*

It is quite permanent, and a highly interesting diminutive fern. The fronds are from about two inches to six inches in length, very variable in form and character; sometimes symmetrical and single, and then pinnate, with oblong obtuse lobate or serrate pinnæ, and a crispy dilated apex. The fronds are, however, frequently unsymmetrical, being here and there depauperated or irregularly developed; sometimes multifid at the apex, and sometimes ramose, branching either from the stipes, or the lower part of the frond. When thus branched, one or both of the branches may become like the single symmetrical fronds, or may more or less closely resemble the irregular formed ones. It is somewhat analogous to the dwarf curled variety of the Lady Fern, but is not nearly so much crisped. It has not, so far as we know, fructified. The plant was found near Buxton, in Derbyshire, by James Schofield, of Rochdale, a botanist in humble life, from whom it appears to have passed into the hands of Messrs. Stansfield, of Todmorden. [Plate XXXVIII.]

25. *subintegra* (M.). This variety, which occurs in the late Mr. Winch's herbarium, belonging to the Linnean Society, is doubtless a form closely allied to *pumila* (14) and *abbreviata* (13). As in them, the fronds have a very narrow lance-shaped outline; they are dwarf, glandular, and merely pinnate; the pinnæ are short and very obtuse, pinnatifid half way down into blunt oblong lobes, not sinuated or crenated merely, as Mr. Newman's figure (*Hist. Brit. Ferns*, 3 ed. 193) indicates. The sori are large, and form a single line on each side the midvein, about equidistant from it and the margin. It is stated by Dr. Johnstone to have been gathered, long since, in abundance, by the Rev. J. Baird, at Ennis, in the county of Clare, Ireland.

The type form of the species is sometimes met with, having the tips of the fronds bifid or multifid (*dichotoma*), but this slight variation is not more than subpermanent.

THE RIGID BUCKLER FERN.

LASTREA RIGIDA.

L. fronds elongate triangular or lanceolate, bipinnate, glandular; pinnæ tapering; pinnules oblong, blunt, lobed, the segments broad rounded, two- to five-toothed, the teeth not spinulose; indusium convex, persistent, fringed with glands. [Plate XXXIX.]

LASTREA RIGIDA, *Presl, Tent. Pterid.* 77. *Deakin, Florigr. Brit.* iv. 99, fig. 1607. *Babington, Man. Brit. Bot.* 4 ed. 422. *Newman, Hist. Brit. Ferns,* 2 ed. 191. *Moore, Handb. Brit. Ferns,* 3 ed. 114; *Id., Ferns of Gt. Brit. Nature Printed,* t. 18. *Sowerby, Ferns of Gt. Brit.* 22, t. 11.

POLYPODIUM RIGIDUM, *Hoffmann, Deutschl. Fl.* ii. 6.

POLYPODIUM FRAGRANS, *Villars, Hist. des Pl. Dauph.* iii. 843; not of Linnæus or Hudson.

POLYPODIUM VILLARSII, *Bellardi, App. Fl. Pedem.* 49.

POLYPODIUM HELIOPTERIS, *Börkhausen, Röm. Archiv. Bot.* i. 19; according to Weber and Mohr.

POLYPODIUM ODORATUM, *Poiret, Encyc. Bot.* v. 541 (excl. syn. Lin. et hab. Sibir.)

ASPIDIUM RIGIDUM, *Swartz, Schrad. Journ. Bot.* 1800, ii. 37; *Id., Syn. Fil.* 53. *Schkuhr, Krypt. Gew.* 40, t. 38. *Hooker, Supp. Eng. Bot.* t. 2724. *Hooker & Arnott, Brit. Fl.* 7 ed. 585. *Bentham, Handb. Brit. Fl.* 630. *Lowe, Nat. Hist. Ferns,* vi. t. 21. *Fries, Sum. Veg.* 82. *Flora Danica,* t. 2187. *Willdenow, Sp. Plant.* v. 265. *Sprengel, Syst. Veg.* iv. 106. *Sturm, Deutschl. Fl.* (Farrn.) t. 2. *Tenore, Att. Accad. del R. Inst. Sc. Nat. Nap.* v. (reprint 20, t. 2, fig. 4). *Fée, Gen. Fil.* 291. *Mettenius, Fil. Hort. Bot. Lips.* 93.

ASPIDIUM FRAGRANS, *Gray, Nat. Arr. Brit. Pl.* 9; not of Swartz.

ASPIDIUM PALLIDUM, *Link, Fil. Sp. Hort. Berol.* 107. *Fée, Gen. Fil.* 291. (*A variety,* β.)

ASPIDIUM NEVADENSE, "*Boissier;*" *Kunze, Ind. Fil. in Linnæa,* xxiii. 229. (*A variety,* β.)

ASPIDIUM ARGUTUM, *Kaulfuss, Enum. Fil.* 242. (*A variety,* γ.)

ASPIDIUM AFFINE, "*Rb.*"; *Visiani, Fl. Dalm.* i. 39.

NEPHRODIUM RIGIDUM, *Desvaux, Ann. Soc. Linn. de Paris,* 261.

POLYSTICHUM RIGIDUM, *De Candolle, Fl. Franç.* 3 ed. ii. 560. *Koch, Synops.* 2 ed. 979. *Ledebour, Fl. Ross.* iv. 516.

POLYSTICHUM STRIGOSUM, *Roth, Fl. Germ.* iii. 86.

LOPHODIUM RIGIDUM, *Newman, Phytol.* 1851, *App.* xxi.; *Id., Hist. Brit. Ferns,* 3 ed. 175.

Caudex thick, scaly, tufted, decumbent, formed of the bases of the decayed fronds closely surrounding a woody axis. *Scales* lanceolate attenuate, and linear-lanceolate, or subulate. *Fibres* long and wiry, branched, dark-coloured.

Vernation circinate.

Stipes short, about one-third of the length of the entire frond, sometimes more; lateral and adherent to the caudex; thickened at the base, glandular; densely clothed with long subulate or linear-lanceolate narrow-pointed membranaceous scales intermixed with broader ones, all of which are of a reddish-brown colour; these scales become smaller and less abundant upwards. *Rachis* furnished with scattered hair-like scales; both primary and secondary rachides bearing numerous short-stalked translucent glands.

Fronds from one to two feet high, firm, dull green, paler beneath, the surface sprinkled over while young with numerous minute spherical short-stalked almost sessile glands, which give it then a glaucous hue, not conspicuous in the dried plants, and at the same time impart a slight but peculiar and agreeable fragrance; they are spreading or erectish, bipinnate, usually elongately triangular, the lower pinnæ being somewhat the longest, and the rest gradually shortening to the apex; sometimes, however, the outline is lanceolate. *Pinnæ* alternate, the lower ones subopposite, distinctly triangular, the middle ones more or less oblong with a tapering point, the uppermost tapering from their base towards their point. *Pinnules* oblong or ovate-oblong, truncate at the base, obtuse at the apex, the lower ones shortly stalked, the upper adnate, deeply pinnatifid; the lobes are oblong notched, the upper with about two, the lower with about five teeth, which are short, and acute but not spinulose.

Venation of the pinnules consisting of a sinuous *costa* or midvein, branching alternately, so as to send out a *vein* into each lobe; these veins branch so as to produce a *venule* for each marginal tooth, towards which it extends, but does not reach the margin; the lower anterior venule is fertile.

Fructification on the back of the frond, occupying about the upper half. *Sori* rather large, round, numerous and occupying the whole length of the pinnules, indusiate, medial on the basal anterior venules, forming a line on each side of and near to the midvein, becoming crowded and often confluent over the whole central portion of the pinnules. *Indusium* lead-coloured, firm, membranaceous, persistent, convex, roundish-reniform, *i. e.*, round with a posterior sinus or notch, by which it is affixed, furnished both on the surface and at the margin with stalked glands. *Spore-cases* numerous, brown, obovate. *Spores* roundish oblong, granulated.

Duration. The caudex is perennial. The fronds are annual, produced in spring and perishing in autumn.

This species may be known from those to which it is most nearly allied by several characteristics. The fronds are comparatively small, and generally broadest at the base, and they are always covered with minute glands, which give off a pleasant balsamic fragrance, often appreciable in the vicinity of the living plants during sunshine. The outline of the pinnules, which are bluntly oblong with shallow lobes, differing in this respect from the other native species of the genus, is most nearly approached by some states of the Incised Male-Fern, and the serratures also, as in that, are not at all spinulose or awn-tipped, but are short and merely acute; it is, however, distinguished from that by its size, its outline, its glandular surface, and its glandular-fringed indusium. It can hardly be mistaken for any other of the *Lastreas*, nearly all the rest of which have spinulose serratures.

This species is local in its range, being almost entirely confined, as far as regards the United Kingdom, to a few limestone craggy mountainous tracts "within a small area in the approximating portions of the counties of Westmoreland, Lancaster, and York." The Rev. G. Pinder writes:*—"I met with *Lastrea rigida* in great profusion along the whole of the great scar limestone district, at intervals between Arnside Knot, where it is comparatively scarce, and Ingleborough, being most abundant on Hutton Roof Crags and Farlton Knot, where it grows in the deep fissures of the natural platform, and occasionally high in the clefts of the rocks; it is generally much shattered by the winds, or cropped by the sheep, which seem to be fond of it. With regard to the shape of the frond, I may mention that among some hundreds of specimens, I

* Newman, *History of British Ferns*, 2d ed. 192.

found but one or two which [had the fronds oblong-lanceolate], all being more or less triangular, and not having the lower pair of pinnæ shorter than those in the upper and middle parts of the fronds. The fronds of young plants are remarkably triangular. The two forms of fronds no doubt depend upon the situation, whether sheltered or otherwise, and on other causes; still I imagine the triangular to be the true form of the plant, having been informed by a person resident in the neighbourhood, that the plant from Ingleborough [with oblong fronds] assumes the triangular form in cultivation. I do not know whether it has been recorded that this fern possesses a slight scent, not at all unpleasant, but strikingly different from that of other ferns." There is a record of a single plant having been found near Bath—probably planted there; and it is reported to have been found quite established, on a wall of clay-slate overhung by trees, at Townley Hall, in the county of Louth, Ireland. In England, its elevation above the sea would appear to range between 1200 feet and 1500 feet, or thereabouts. The actual habitats on record are:—

Peninsula.—Cornwall, *Hb. Hooker.* Somersetshire: Bath, probably planted.

Mersey.—? Lancashire: Woolston Moss, near Warrington.

Humber.—Yorkshire: Wharnside; White Scars, above Ingleton, on the north side of the valley; also at the foot of Ingleborough, on the north-west side; at the foot of Attermine Rocks, near Settle, at about 1550 feet above the sea.

Lakes.—Westmoreland: Arnside Knot; Hutton Roof Crags; Farlton Knot. North Lancashire: Silverdale; near the top lock of the Lancaster and Kendal Canal.

Leinster.—Louth: Clay-slate wall at Townley Hall, probably introduced, *C. L. Darby.*

This Fern is spread over the Alpine districts of the middle and south of Europe, extending northwards to Norway. It is recorded from France, Switzerland, Mount Cenis in Sardinia, Naples, Calabria, Sicily, Germany, Dalmatia, Croatia, and Hungary. It has been obtained from Mount Sypilos in Asia Minor, from Imeritia in

Asiatic Russia, and from Siberia. The variety *pallida,* which is scarcely different, is found in the south of Europe where it spreads from Italy and Spain, to Greece, as well as in Northern Africa. The *Aspidium argutum* of Kaulfuss, from California, is merely another variety, larger and more developed. The *Dryoptera rigida* of Dr. Asa Gray, found, though rarely, in Massachusetts, North America, which acquires a larger size and more developed character than the European *rigida,* proves to belong to the *spinulosa* type.

The culture of this Fern is very similar to that of the other larger growing kinds, except that it is more impatient of moisture, and doubtless prefers a purer atmosphere. It grows well in free well-drained loamy soil; and the fact of its range being almost if not quite confined to limestone mountains, suggests that the use of limestone among the soil may be beneficial, though it is certainly not essential to success. It is of far more importance that the soil should be kept moderately moist, and should be of such a texture, as may at the same time prevent any accumulation of stagnant water. It is increased by separating the lateral crowns formed by the caudex. The latter is the better for being somewhat elevated above the soil in planting, for being decumbent in habit, it does not when planted deeply, liberate its crown so readily as the more erect-habited species.

Mr. Wollaston has noticed a variation in which the fronds or the pinnæ, or both, are simply or multifidly divided at the apex; and Messrs. Stansfield and Son, of Todmorden, have a small ramose form, obtained from Ingleton Fells, but neither of these prove to be constant forms.

THE CRESTED BUCKLER FERN.

LASTREA CRISTATA.

L. fronds erect, narrow linear-oblong or oblong-lanceolate, sub-bipinnate or bipinnate; serratures spinose-mucronate; scales of the stipes ovate, pallid, scattered; indusium without marginal glands.

—(type): fronds narrow linear-oblong; pinnæ short triangular; pinnules or segments oblong, nearly always connected at the base, crenate-serrate, or obscurely lobed, anterior and posterior ones of the lower pinnæ nearly equal. [Plate XL.]

LASTREA CRISTATA, *Presl, Tent. Pterid.* 77, t. 2, fig. 10. *Hooker, Gen. Fil.* t. 45 A, fig. 4-9. *Deakin, Florigr. Brit.* iv. 107, fig. 1610. *Moore, Handb. Brit. Ferns,* 3 ed. 117; *Id., Ferns of Gt. Brit. Nature Printed,* t. 19. *Babington, Man. Brit. Bot.* 4 ed. 421. *Newman, Hist. Brit. Ferns,* 2 ed. 203. *Sowerby, Ferns of Gt. Brit.* 21, t. 10.

LASTREA CALLIPTERIS, *Newman, Hist. Brit. Ferns,* 2 ed. 12.

POLYPODIUM CRISTATUM, *Linnæus, Sp. Plant.* 1551; according to the Linnæan Herbarium.

POLYPODIUM CALLIPTERIS, *Ehrhart, Beitr. zur Naturk.* iii. 77.

ASPIDIUM CRISTATUM, *Swartz, Schrad. Journ. Bot.* 1800, ii. 37; *Id., Syn. Fil.* 52. *Smith, Eng. Bot.* xxx. t. 2125 (not xxviii. t. 1949, which is *L. Filix-mas*); *Id., Eng. Fl.* 2 ed. iv. 276. *Hooker & Arnott, Brit. Fl.* 7 ed. 584. *Bentham, Handb. Brit. Fl.* 630. *Schkuhr, Krypt. Gew.* 39, t. 37. *Willdenow, Sp. Plant.* v. 252. *Sprengel, Syst. Veg.* iv. 104. *Fries, Sum. Veg.* 82. *Hooker, Fl. Lond.* iv. t. 113. *Svensk Bot.* t. 390. *Fée, Gen. Fil.* 291. *Mettenius, Fil. Hort. Bot. Lips.* 93. *A. Gray, Man. Bot. North. U. States,* 2 ed. 598. *Lowe, Nat. Hist. Ferns,* vi. t. 20.

ASPIDIUM CRISTATUM, β. CALLIPTERIS, *Pursh, Fl. Amer. Sept.* ii. 662.

ASPIDIUM GOLDIANUM, *of some gardens;* not of Hooker and Greville.

ASPIDIUM LANCASTRIENSE, *Sprengel, Anleit.* iii. 134. *Schkuhr, Krypt. Gew.* 41, t. 44. *Willdenow, Sp. Plant.* v. 261. *Sprengel, Syst. Veg.* iv. 104. (*A variety,* β.)

ACROSTICHUM CALLIPTERIS, *Ehrhart;* according to Sadler.

NEPHRODIUM CRISTATUM, *Michaux, Fl. Bor. Amer.* ii. 269.

POLYSTICHUM CRISTATUM, *Roth, Fl. Germ.* iii. 84. *Koch, Synops.* 2 ed. 978. *Ledebour, Fl. Ross.* iv. 515.

POLYSTICHUM CALLIPTERIS, *De Candolle, Fl. Franç.* 3 ed. ii. 562.

DRYOPTERIS CRISTATA, *A. Gray, Man. Bot. North. U. States,* 1 ed. 631.

LOPHODIUM CALLIPTERIS, *Newman, Phytol.* iv. 371; *Id.,* 1851, *App.* xix.; *Id., Hist. Brit. Ferns,* 3 ed. 169 (excl. syn. Hoffm.).

Var. uliginosa: fronds—earlier fertile ones tall, erect, narrow, linear-lanceolate, bipinnate below; the pinnules oblong-acute, mostly adnate, inciso-serrate or lobed, with aristate incurved teeth; barren ones shorter, their pinnules oblong bluntish, adnate or decurrent, crenate-serrate; later fertile ones broader than the earlier, the pinnules oblong bluntish crenate-serrate as in barren ones; anterior and posterior basal pinnules of the lowest pinnæ in all the fronds nearly equal in size. [Plate XLI.]

Lastrea cristata, β. uliginosa, *Moore, Trans. Bot. Soc. Edin.* iv. 109; *Id., Phytol.* iv. 149; *Id., Handb. Brit. Ferns*, 3 ed. 117; *Id., Ferns of Gt. Brit. Nature Printed*, t. 20. *Babington, Man. Brit. Bot.* 4 ed. 422.
Lastrea uliginosa, *Newman, Phytol.* iii. 679.
Aspidium cristatum, β. uliginosum, *Hooker & Arnott, Brit. Fl.* 7 ed. 585.
Aspidium spinulosum, *Hooker & Arnott, Brit. Fl.* 6 ed. 571, in part.
Aspidium spinulosum, *b.* uliginosum, *A. Braun: Döll, Rhein. Fl.* 17.
Aspidium spinulosum, *v.* subcoriaceum, *Ruprecht, Dist. Crypt. Ross.* 37.
Aspidium cristatum × spinulosum, *Milde, Hb. Hook.*
Polystichum spinulosum, *v.* uliginosum, "*A. Br.*": *Wirtzen, Crypt. Vasc. Rhen. Pruss;* according to Newman.
Lophodium uliginosum, *Newman, Phytol.* iv. 371; *Id.*, 1851, *App.* xix.; *Id., Hist. Brit. Ferns*, 3 ed. 163.

Var. **spinulosa**: fronds narrow oblong-lanceolate, bipinnate; pinnæ triangular, oblique; pinnules oblong, acute, inciso-serrate or pinnatifid, with aristately toothed lobes; posterior basal pinnules of the lower pinnæ much larger than the anterior ones. [Plate XLII.]

Lastrea cristata, γ. spinulosa, *Moore, Handb. Brit. Ferns*, 2 ed. 115; 3 ed. 117; *Id., Ferns of Gt. Brit. Nature Printed*, under t. 19.
Lastrea spinulosa, *Presl, Tent. Pterid.* 76. *Babington, Man. Brit. Bot.* 4 ed. 422. *Sowerby, Ferns of Gt. Brit.* 24, t. 12. *Moore, Ferns of Gt. Brit. Nature Printed*, t. 21.
Lastrea spinosa, *Newman, Nat. Alm.* 1844, 21. *Deakin, Florigr. Brit.* iv. 108, fig. 1611.
Lastrea dilatata, *v.* linearis, *Babington, Man. Brit. Bot.* 1 ed. 386 (excl. syn.)
Polypodium spinulosum, *Müller, Fl. Fridrichsdal.* 193, n. 841, t. 2, fig. 2.; *Id., Fl. Dan.* iv. fasc. 12, 5, t. 707.
Polypodium cristatum, *Hoffmann, Deutschl. Fl.* ii. 8 (excl. syn. Bolt.)
Polypodium multiflorum, β. spinosum, *Roth, Catal. Bot.* i. 141.
Polypodium Filix-fœmina, *v.* spinosa, *Weis, Pl. Crypt.* 316.
Aspidium spinulosum, *Swartz, Syn. Fil.* 420 (not 54, nor Schrad. Journ.). *Schkuhr, Krypt. Gew.* 48, t. 48 (excl. fig. *d, e*). *Sprengel, Syst. Veg.* iv. 106 (in part, *i.e.*, excl. character of indusium, and synonym). *? Bentham, Handb. Brit. Fl.* 630, in part. *Fée, Gen. Fil.* 291. *Mettenius, Fil. Hort. Bot. Lips.* 93. *Lowe, Nat. Hist. Ferns*, vi. t. 41.
Aspidium spinulosum, *a.* elevatum, *A. Braun: Doll. Rhein. Fl.* 17.
Aspidium spinulosum-cristatum, *Laschner, Bot. Zeit.* xiv. 435.
Aspidium spinulosum, β., *Ruprecht, Dist. Crypt. Ross.* 37.
Nephrodium spinulosum, *Strempel, Fil. Berol. Syn.* 30.
Polystichum spinosum, *Roth, Fl. Germ.* iii. 91.
Lophodium spinosum, *Newman, Phytol.* iv. 371; *Id.*, 1851, *App.* xviii.; *Id., Hist. Brit. Ferns*, 3 ed. 157.

Caudex stoutish, decumbent or slowly creeping, *i. e.*, extending in a horizontal direction, the fronds of each season being in advance of those of the preceding year; branched, scarcely tufted, somewhat scaly, formed of the enlarged living bases of the decayed fronds surrounding a woody axis. *Scales* similar to those of the stipes. *Fibres* numerous, coarse, dark brown, branched.

Vernation circinate, the pinnæ lying flat against the sides of the incurved rachis.

Stipes terminal and adherent to the caudex, about one-third of the entire length of the frond, stout, shining, dark brown at the base, the brown blending with green upwards, sparsely scaly, with broad ovate membranaceous pale-brown scales, which are for the most part appressed, and are most numerous near the base. *Rachis* stout, channelled in front, almost free from scales, pale green.

Fronds from one to three feet high, herbaceous, dull green, erect, narrrow linear-oblong, tapering at the apex, scarcely at all narrowed at the base, subbipinnate. *Pinnæ* numerous, the lower ones distant, subopposite, short triangular, two inches long, an inch and a half broad at the base; the upper more contiguous, alternate, elongate triangular, those near the middle of the frond measuring about two and a half inches long, and nearly an inch and a quarter broad at the base; all shortly stalked, the stalk twisted so that their upper surface is directed towards the apex of the frond. *Pinnules* oblong, bluntish, more or less adnate, and connected by the wing of the rachis, the basal ones only, and these only on highly developed fronds, having a narrow attachment, pinnatifidly lobed, the lobes serrate, with spinulose teeth; the rest of the pinnules are inciso-crenate at the margin, serrate at the apex, the crenatures serrated, and all the serratures tipped by a spinulose point. The posterior basal pinnules are scarcely larger than the anterior ones of the same pinnæ; while those of the late summer and autumnal fronds are broader and larger.

Venation of the pinnules consisting of a flexuous *costa* or midvein, which throws off a *vein* into each lobe; these veins bear several *venules*, which are either simple or forked, and are directed one towards each tooth, terminating within the margin in a somewhat thickened point. Usually only the anterior basal venule of each fascicle bears a sorus, but occasionally on the lowest pinnules the posterior basal venule also is fertile. The veins are conspicuously depressed on the upper surface.

Fructification on the back of the fronds, usually confined to the upper half, but sometimes extending lower down. *Sori* numerous, round, indusiate, medial on the anterior basal venules, in a row on each side of and nearer to the midrib than the margin, except in the most luxuriant pinnules, where the development of sori on the posterior venules produces a more irregular arrangement. *Indusium* membranaceous, reniform, flat, with a wavy, somewhat irregular margin, but without glands, affixed by a deep basal sinus. *Spore-cases* numerous, dark brown, roundish. *Spores* oblong, granulated.

Duration. The caudex is perennial. The fronds are annual, the earliest being produced in May, and these are succeeded by others during the summer, all becoming destroyed by the autumnal frosts, or perishing even if not exposed.

Lastrea cristata, with the plants called *uliginosa* and *spinulosa*, form a group distinguishable by habit and other characters from the allied *dilatata* group, with which, however, the more highly developed form, *spinulosa*, is sometimes associated by botanists of high authority—in consequence, no doubt, of the plants having been studied in the herbarium, where their differences become less marked, rather than in a state of growth, in which certain important characters are obvious. Of this first-mentioned group, *Lastrea cristata* is the least developed form. In our *Handbook of British Ferns* (2 ed.) it was treated as consisting of three forms of one not very variable species; and notwithstanding that many Fern authorities do not appear to adopt this view, we have no doubt whatever that the plants possess a close natural affinity, and have characters which separate them from the forms of *Lastrea dilatata*, however similar to the latter, in some cases, may be the degree and mode of division in the fronds—points on which botanists are at times too prone to rely. The close affinity of the three forms now alluded to, is evidenced by marks far more important than those to be derived from such characters as the outline or cutting of the fronds: namely, by the creeping caudex, by the erect narrow fronds, by the sparse and pallid broad appressed scales of the stipes, and by the entire indusia, in all which respects they perfectly agree. On the other hand, it is in these points that they differ from the *dilatata* group. In the folio edition of this work, we were led, in deference to the more commonly received opinion, to treat of *spinulosa* separately, but after some years' further observation we revert to our former view, and place it here under *cristata*.

The plant which we regard as the type of this group is rare and local in this country; occurring only in boggy situations. The counties of Norfolk and Suffolk, on the eastern side of England, seem its head quarters, and thence it stretches westwards to those of Nottingham and Chester. Some others, as Huntingdonshire and Staffordshire, have also been reported, and are sufficiently probable; but these, together with some other English as well as Scotch and Irish habitats, either require confirmation, or are altogether erroneous. Mr. Watson estimates its range in altitude as extending from the sea level, to about 300 feet above it. The variety *uliginosa* is limited in its range; *spinulosa* is more generally distributed. The distribution of the species is recorded as follows:—

Ouse.—Suffolk: Westleton, *Eng. Fl.*; Bexley Decoy, near Ipswich, *H. Bidwell.* Norfolk: Bawsey Heath, near Lynn; Dersingham; Edgefield, near Holt; Holt Lows, *Rev. W. H. Girdlestone;* Fritton, near Yarmouth; Surlingham Broad, near Norwich, *Rev. W. S. Hore;* Wymondham; Fakenham, *W. G. Johnstone.* ? Huntingdonshire. ? Bedfordshire.

Severn.—Staffordshire: near Madeley; Bog near Newcastle-under-Lyne, *J. Hardy, Hb. Leighton.* ? Worcestershire.

Trent.—Nottinghamshire: Oxton Bogs, *Dr. Howitt;* Bullwell Marshes.

Mersey.—Cheshire: Wybunbury Bog, *Rev. G. Pinder.*

Humber.—Yorkshire: Plumpton Rocks, near Knaresborough, *Baines's Flora;* Malton, *J. Mackell, C. Monkman.*

This Fern is a generally dispersed European species, occurring from the Scandinavian kingdoms to Moscow on the one hand, and also in Holland, Belgium, Germany, the Carpathians, Hungary, Transylvania, Croatia, France, Switzerland, Italy and Bœotia. It is said to grow in the Caucasian range, and in Siberia. It also occurs in North America, both in the United States and in Canada, where occur two or three forms, one of which is the *Aspidium lancastriense*, and another, common in gardens, is very generally mistaken for *Aspidium Goldianum*. It has also been obtained from the Slave River, in North West America. The variety *uliginosa* occurs in Germany as well as in England; and in Dahuria. The variety *spinulosa* is probably not uncommon, but its range is not at all accurately known, in consequence chiefly of the confusion which has generally existed between this plant and *L. dilatata*, which renders almost all the published statements open to doubt. The fragmentary condition, too, of many of the foreign specimens preserved in herbaria, renders it impossible to employ to the full extent this source of information. The following countries and habitats may, we believe, be confidently cited:—Denmark; Sweden; St. Petersburgh and Moscow, in Russia; Great Britain and Ireland; France, Switzerland; Germany in various parts; and Hungary. We believe we may also here refer specimens in the Hookerian herbarium, from Labrador, Boston, and Canada; though, according to Dr. Asa Gray, the common American plant of this affinity is not *spinulosa* but *intermedia*.

These Ferns grow readily in peaty soil, with abundant moisture; and though not remarkable for elegance, they have a certain distinctness of character, and are useful in grouping on account of their upright habit of growth. They are increased with tolerable facility by the separation of the lateral crowns which are frequently produced.

The species and its varieties produce occasional multifid variations, sometimes consisting in the division of the apices of the pinnæ, sometimes in that of the apex of the frond, but these forms cannot rank as permanent varieties, like the following:—

1. *uliginosa* (M.). This plant, the *Lastrea uliginosa* of Newman, we regard as a variety of *cristata*, the only marked difference, in

truth, being that its earlier fertile fronds have the pinnules more acute, and more conspicuously lobed and toothed, and that the discrepancy in size between the anterior and posterior basal pinnules of the lower pinnæ, is a trifle more manifest. *Caudex* stout decumbent, branching sparingly. *Fronds* erect linear-lanceolate, from two to four feet high, bipinnate at the base of the pinnæ; of three kinds, but not all simultaneous in their appearance, nor constantly produced, and though different not strikingly dissimilar like the barren and fertile fronds of some other Ferns:—(1). The spring or early fronds of strong crowns, which are fertile; (2) with these sometimes, but not always, appear others which are shorter and barren, the latter being often produced from small lateral crowns, but also sometimes from the same crown which produces the fertile ones; (3) summer or later fronds, which have broader and blunter pinnules more like the early barren ones, and being sometimes fertile, and sometimes barren. The early barren fronds (2) are small, spreading, pinnate, with decurrent oblong obtuse pinnules, and resemble small barren fronds of *cristata*. The summer fronds (3) are also like *cristata*, but perhaps broader, with decurrent oblong obtuse pinnules, and they are very frequently fertile. The fertile fronds produced in spring (1) are more like *spinulosa*; they grow quite erect, and are linear-lanceolate, bipinnate, the basal pinnules distinct, all the pinnæ stalked, and set on so that their upper surface is turned towards the point of the frond. *Pinnæ* elongate triangular, the lower ones being shorter, broader, and more oblique than those above them, their first posterior pinnule being an inch, and the opposite anterior pinnule three-fourths of an inch long. The basal *pinnules* of the middle pinnæ are distinct, oblong, acute, pinnatifidly lobed, the lobes sharply serrate with longish spinulose or aristate teeth; the upper pinnules are adnate and sharply and deeply serrate. *Fructification* extending over the whole frond, but most copious towards the top, where it forms two lines near the midrib on the smaller pinnules, being confined to the anterior basal venules; whilst it becomes confused on the larger pinnules in consequence of being produced in two series on the lobes, both anterior and posterior venules being there fertile. The fronds are said to appear several days earlier than those of *Lastrea cristata*, but our cultivated plants of both have never shown any constancy in this respect, although circumstanced alike. We consider this plant more closely allied to the type *cristata* than to *spinulosa,* because its vernation agrees more exactly with the former, and because neither its early barren fronds nor its later fertile ones can be certainly distinguished from analogous separated fronds of *cristata,* whilst, on the other hand, no such *cristata*-like fronds are produced by *spinulosa.* The special attention of English botanists was drawn to this plant a few years since by Mr. John Lloyd, by whom it was found at Oxton Bog, Nottinghamshire. It has also been certainly gathered at Wybunbury Bog, Cheshire; and at Bawsey Heath, Norfolk. There are beyond this records of its occurrence at Surlingham Broad and Wymondham, both in Norfolk; and we have seen specimens which are, we think, referrible here from Tunbridge Wells in Kent, from a bog near Newcastle-under-Lyne, in Staffordshire; and from Broseley in Shropshire. Mr. Newman himself reports it from Epping Forest, but all the so-called *uliginosa* we have seen from thence proved to be *spinulosa.* From Scotland there is no recorded station. In Ireland, Mucruss, Killarney, has been mentioned as a habitat, and Wager, a Fern-collector, reports that it is plentiful in Glen Flesk, near Kenmare, in Kerry. The same plant doubtless occurs in Rhenish Prussia.

2. *spinulosa* (M.). This plant is variously considered as a form of *Lastrea dilatata,* as a distinct species, and as a form of *Lastrea cristata* which latter is the view we adopt. It is distinguished from *L. dilatata,* by its creeping caudex, by the few broad pallid scales of its stipes, and by the absence of glands from the margin of its indusium. The connecting link between it and *dilatata* is the *glandulosa* of Newman, which latter has a decumbent and in some instances a slowly creeping caudex, but not the entire margined indusium of *spinulosa,* and differs also in the abundant glands which cover it—though this latter is a character of comparatively little value, for common forms of *dilatata,* in no other respect distinguishable, are found both covered with and free from glands. Though agreeing with *cristata* in the precise and import nt characters afforded by the caudex, the scales, and the indusia, it is perhaps more readily separated by the eye from that, than from imperfect specimens of

dilatata, with some forms of which it accords in the variable, and therefore less important character of the subdivision of its parts. *Lastrea cristata* itself may be separated from *spinulosa* by its short triangular, and less divided pinnæ, and by their blunter, less deeply toothed pinnules; but the variety *uliginosa* is in some of its states much less easily known from it, the greater inequality of the pinnules on the lower pinnæ of *spinulosa* being almost the only difference; indeed so closely do these merge into each other by means of transition forms of frond, that we are forced to the conclusion that all three—*cristata*, *uliginosa*, and *spinulosa*, are in reality, as we place them, variations of one specific type.

There are two versions of the name of this plant in use among British botanists—*spinosa* and *spinulosa*. We advisedly use the latter. The former has been revived by recent authors, on the grounds that Roth, who employed it in *Flora Germanica* (1800), was the first to correctly define the plant from the allied *dilatata*, and that Müller in the *Flora Danica* has 'misprinted' *spinulosa* for Weis' name of *spinosa*, and under it figured the plant we now call *Lastrea cristata*. Now, Weis' name *Polypodium Filix-fœmina* var. *spinosa*, as that of a variety merely, and altogether so incorrect as to the species, has no claim to notice; and Müller describes, but without name, and very well figures, two pinnæ of *spinulosa* in the *Flora Fridrichsdalina* (1767), and his later figure in *Flora Danica* (1777), where he names it *Polypodium spinulosum*, is an exact representation of our *spinulosa*, and not of the species *cristata*. We therefore can neither subscribe to the assumption that Müller's name is a misprint, nor allow the claim made on behalf of Weis' name; while Müller has by many years the precedence over Roth. Equally, as we believe, are those writers in error, who deny that this plant is the *Lastrea spinulosa* of Presl; for *Aspidium spinulosum* as defined by Swartz in his *Synopsis Filicum* (p. 420) is the plant of Müller's figures above referred to; and Swartz moreover quotes Schkuhr's t. 48, which perfectly represents *spinulosa*, excepting, however, the detached figures of indusia, *d* and *e*, these latter being, as we believe, erroneous, glandular indusia not having been found on the true *spinulosa*. Swartz's plant, therefore, is our *spinulosa*, not *dilatata*, and that of Presl is the same species with a new generic name. We append a more complete description of this plant, than of our other varieties, for the convenience of those who still regard it as a distinct species:—

Caudex stoutish, decumbent or slowly creeping in a horizontal direction, with the fronds growing erect from its apex; branched, sometimes more or less tufted, slightly scaly, formed of the enlarged and enduring bases of the decayed fronds, surrounding a woody axis. *Scales* resembling those of the stipes. *Fibres* coarse, numerous, branched, dark brown.

Vernation circinate; sometimes in this plant the rachis is simply circinate, but in other cases besides the ordinary involution, there is also a lateral curvature; the pinnæ and pinnules are all separately involute.

Stipes terminal and adherent to the caudex, nearly as long as the leafy part of the frond, stoutish, dark brown-purple at the base; sparsely scaly, with broad-ovate membranaceous pale-brown scales of which many become at length more or less appressed; the scales are most numerous near the base. *Rachis* stoutish, channelled in front, scarcely at all scaly, pale green, smooth.

Fronds from two to four or five feet in height, erect, herbaceous, yellowish green, narrow oblong-lanceolate tapering at the apex, bipinnate. *Pinnæ* numerous, opposite or subopposite below, often becoming more alternate above; the lower ones distant, obliquely triangular, from the greater size of the posterior basal pinnules, measuring (in average specimens, two feet or upwards in height) about four inches in length, and three inches across the base, of which latter the posterior pinnules measure nearly two inches; the upper ones are less distant and narrower, of an elongate triangular outline, those just above the middle measuring four and a half inches long, and barely two inches broad at the base, where the posterior and anterior pinnules are of nearly equal size. The pinnæ are stalked, frequently more or less drooping, and often twisted so as to turn their upper surface towards the apex of the frond, but this peculiar twisting is less marked than in *cristata*. *Pinnules* oblong acute, broadest at the base, the lower ones with a short stalk-like attachment, the upper more or less adnate; the basal pinnules (of the pinnæ half-way up the fertile fronds) pinnatifid almost to the midrib, with oblong acute lobes; the lobes strongly serrated, with spinulose teeth, whose points are directed towards the apex of the lobe, and often curved upwards above the plane of its surface; the upper pinnules are either inciso-lobate with spinulosely serrate lobes, or coarsely serrate with spinulose teeth. The barren fronds usually, and some of the fertile ones, are broader and more lax in habit than those above described, and sometimes entire plants assume this character.

Venation of the pinnules (the basal ones of fertile pinnæ near the centre of the frond,) consisting of a stout midvein, from which a primary vein extends into each lobe, where it forms a flexuous secondary midvein, bearing alternate forked *venules*, on the short anterior fork of which, nearly at its point, and standing just beneath the sinus of the serrature, the sorus is placed, the sori then forming two rows along the lobes of the pinnules. In the less divided pinnules at the middle of the pinnæ, the primary midvein produces branched *veins*, and the anterior basal *venule* also in this case bears the sorus, near to its termination, so that the sori then form two lines along the pinnule itself. This latter being the structure of the greater number of pinnules, the general aspect of the fructification is to form two lines lengthwise on the pinnule. The venules are directed one towards each serrature, but terminate before reaching it, in a thickened point.

Fructification on the back of the frond, usually occurring on the upper half, but sometimes extending over the whole surface. *Sori* numerous, round, indusiate, medial or subterminal on the anterior basal venules, (or on several venules in the deeply pinnatifid basal pinnules,) forming a line on each side the midvein; usually distinct, but often crowded. *Indusium* flat, reniform, membranaceous, persistent, with an entire margin, wavy or with angular projections, but without glands. *Spore-cases* brown, numerous, rotundate. *Spores* oblong, granulated.

Duration. The caudex is perennial. The fronds are annual, the first growth appearing early in May, and others growing up at intervals through the summer; they perish in autumn when exposed, but under shelter, though decaying near the base of the stipes so as to be unable to stand erect, they nevertheless retain much of their freshness through the winter; and the extreme base of the stipes continues fresh for many years.

This variety, by many considered as a species, is common, and doubtless generally distributed, though not always well discriminated from *Lastrea dilatata;* the records of its distribution are, consequently, not free from doubt. It however grows generally in damp shady places, and occurs over the whole of England. In Wales it seems less general, and is there most plentiful in the southern parts. In Scotland and in Ireland it is apparently rare. Its range in elevation appears to be between the coast level, and an altitude of about 600 feet in England, or probably considerably more in Scotland. The following are the habitats:—

Peninsula.—Cornwall: about Penzance. Devonshire: Fingal Bridge; Exwick Wood, near Exeter, *R. J. Gray;* Eggesford, *H. F. Dempster;* Barnstaple, *C. Jackson.* Somersetshire: Selworthy, and elsewhere.

Channel.—Hampshire. Isle of Wight: Tinker's Hole, Apse Castle, and elsewhere. Dorsetshire. Sussex: Ardingly, *F. Evans;* Woolbedding, near Midhurst, *F. Bourdillon;* Tilgate Forest, *J. Lloyd;* Tunbridge Wells.

Thames.—Hertfordshire: Ball's Wood, Hertford; North Mimms; Hatfield, etc. Kent: Chislehurst; Canterbury, etc. Middlesex. Surrey: Gomshall; Combe Wood; Wimbledon; Barnes, *T. M.;* Portnall Park, Virginia Water, *T. M.*, etc. Buckinghamshire: Fulmer, *J. Lloyd;* Black Park, *Dr. Allchin.* Essex: Epping; Danbury; Coggeshall; Pod's Wood, Tiptree, *E. Hall;* Kavanagh Wood, Brentwood, *S. F. Gray.*

Ouse.—Suffolk. Norfolk: Surlingham Broad, near Norwich; Scoulton Mere, *G. J. Chester;* Bawsey, near Lynn, *Dr. Allchin;* Loddon, *Rev. J. J. Smith;* Holt Lows, *Rev. W. H. Girdlestone,* etc. Cambridgeshire: Fulbourne. Northamptonshire.

Severn.—Warwickshire: North Wood, Arbury Hall; Binley; Rugby; Chesterton Wood; Rounsel-lane, *T. Kirk.* Gloucestershire. Monmouthshire: Pont-y-pool, *T. H. Thomas.* Herefordshire: The Horls, near Ross, *W. H. Purchas.* Worcestershire: Bewdley, *Miss Hall.* Staffordshire: Needwood; Wolverhampton.

Shropshire: Whitchurch, *R. W. Rawson;* Bomere Pool and Shomere Moss, near Shrewsbury, *Rev. W. A. Leighton;* Shirlet, near Broseley, *G. Maw.*

Trent.—Nottinghamshire: Paplewick; Oxton Bogs; Wollaton, *E. J. Lowe.* Leicestershire: Netherscall, *Rev. A. Bloxam.* Derbyshire, *R. M. Norman.*

Mersey.—Cheshire: Wybunbury Moss, near Nantwich, *R. W. Rawson;* Ashton Moss, *H. Buckley;* Delamere Forest. Lancashire: Chat Moss; Lowgill, *H. Shepherd;* Woolston Moss, *T. G. Rylands,* and Risley Moss, near Warrington; Levenshulme, *S. F. Gray.*

Humber.—Yorkshire: Sheffield; Richmond; Ingleborough; Doncaster; Leckby Car; Terrington Car; York; Kildale, *W. Mudd;* Hackness, *A. Clapham;* Thirsk.

Tyne.—Durham: Waldridge Fell, near Durham, *J. Mitchinson;* Northumberland: Chivington Wood, *Rev. R. Taylor.*.

Lakes.—Cumberland: Red-house; Keswick, *F. J. A. Hort.* Westmoreland: Windermere, *F. Clowes.* Isle of Man.

S. Wales.—Brecknockshire. Glamorganshire. Carmarthenshire.

N. Wales.—Carnarvonshire.

W. Lowlands.—? Dumfriesshire, *P. Gray.*

E. Lowlands—? Edinburghshire.

E. Highlands.—Forfarshire: Rescobie, *A. Croall;* Guthrie Woods, *A. Croall.* Perthshire: Dunkeld, *A. Tait;* and elsewhere, *A. Croall.*

W. Highlands.—? Argyleshire. Dumbartonshire: Tarbet, *T. M.*

N. Highlands.—Ross-shire: Brahan Castle, Dingwall, *Sir W. C. Trevelyan, Bart., Miss Murray.*

W. Isles.—North Uist. Harris. Lewis.

Ulster.—Monaghan: Dartrey, *C. L. Darby.* Londonderry: near Dungiven, *D. Moore.*

Connaught.—Galway: Connemara.

Leinster.—Wicklow: Newtown, Mt. Kennedy, *R. Barrington,* and elsewhere, *D. Moore.* Westmeath: Mullingar; and not unfrequent in Ireland, *D. Moore.*

Munster.—Kerry: Killarney, *E. J. Lowe.*

3. *strigosa* (M.). This is a form of the *spinulosa* type, and is remarkable chiefly for the spiny-teeth of the margins of the pinnules

being longer and more bristly than usual. When originally found its pinnules were somewhat depaupérated, distant and confluent, but these features have not been permanent. It was found by Mr. R. Sim, at Chislehurst, Kent.

4. *nana* (Sim). This also belongs to the *spinulosa* type, and is described by Mr. Sim as being about six inches high, and constant to its dwarf pigmy character.

5. *crispa* (M.). This also belongs to the *spinulosa* type, and has been recorded by Dr. Deakin, in the *Florigraphia Britannica* (iv. 108, figs. *e, f,* on p. 111). It is described as being very rigid, the margins of the lobes of the pinnules closely rolled back, and partly concealing the sori; they are said to be so rigidly curled back as to be retained flat with great difficulty when unrolled. The sori are larger and much darker than in the usual forms of *spinulosa.* It was found near Sheffield in Yorkshire.

6. *interrupta* (M.). A slight variation of the *spinulosa* type, in which the pinnules are irregularly developed; here and there depauperated, and in some parts enlarged and broader than usual. It was found at Malton, Yorkshire, by Mr. Monkman.

7. *tripinnata* (M.). A very elegant form of the *spinulosa* type. The fronds are narrow, stiff, erect; the pinnæ short, and the pinnules small, but the anterior and posterior ones unequal as in the most marked forms of *spinulosa;* the pinnules are, however, distinctly divided into little oblong obtuse lobate-serrate or biserrate pinnulets. The general aspect of the frond, is that of being made up of a multitude of little parts. It was purchased of a Fern hawker by Mr. F. C. Wilson, of Stamford Hill.

THE BROAD PRICKLY-TOOTHED BUCKLER FERN.

LASTREA DILATATA.

L. fronds ovate, subtriangular, or oblong-lanceolate, bipinnate, with the pinnules pinnate or pinnatifid, spinosely mucronate-serrate; scales of the stipes numerous, lanceolate, entire or fimbriate, usually dark-centred; indusium fringed with stalked glands.

— (type): fronds ample, ovate, bi-tripinnate; scales of the stipes entire, strongly two-coloured, *i. e.*, with a dark centre and paler margins; indusium prominent. [Plate XLIII.]

LASTREA DILATATA, *Presl, Tent. Pterid.* 77. *Newman, Nat. Alm.* 1844, 23. *Babington, Man. Brit. Bot.* 4 ed. 422. *Sowerby, Ferns of Gt. Brit.* 25, t. 13. *Moore, Handb. Brit. Ferns,* 3 ed. 124; *Id., Ferns of Gt. Brit. Nature Printed,* t. 22.

LASTREA MULTIFLORA, *Newman, Hist. Brit. Ferns,* 2 ed. 215. *Deakin, Florigr. Brit.* iv. 113, fig. 1613.

ASPIDIUM DILATATUM, *Smith, Fl. Brit.* 1125; *Id., Eng. Fl.* 2 ed. iv. 280; *Id., Eng. Bot.* xxi. t. 1461. *Swartz, Syn. Fil.* 420. *Fée, Gen. Fil.* 291. *Mettenius, Fil. Hort. Bot. Lips.* 93. *Tenore, Att. Accad. del R. Inst. Sc. Nat. Napol.* v. (reprint 18, t. 2, fig 3). *Lowe, Nat. Hist. Ferns,* vi. t. 27.

ASPIDIUM SPINULOSUM, *Swartz, Schrad. Journ. Bot.* 1800, ii. 38, in part; *Id., Syn. Fil.* 54, in part, 420. *Smith, Fl. Brit.* 1124; *Id., Eng. Bot.* xxi. t. 1460; *Id., Eng. Fl.* 2 ed. iv. 279, in part. *Willd. Sp. Plant.* v. 262. *Hooker & Arnott, Brit. Fl.* 7 ed. 586 (excl. γ.) *Bentham, Handb. Brit. Fl.* 630, in part. *Fries, Sum. Veg.* 82.

ASPIDIUM SPINULOSUM, *v.* DILATATUM, *Link, Fil. Sp.* 106. *A. Gray, Man. Bot. North. U. States,* 2 ed. 597.

ASPIDIUM SPINULOSUM, δ., *Ruprecht, Dist. Crypt. Ross.* 38.

ASPIDIUM CRISTATUM, *Fl. Wett.*; according to Steudel.

ASPIDIUM CARTHUSIANUM, *Steudel, Nomencl. Bot.* 61.

POLYPODIUM DILATATUM, *Hoffmann, Deutschl. Fl.* ii. 7.

POLYPODIUM CRISTATUM, *Hudson, Fl. Ang.* 457. *Hoffmann, Deutschl. Fl.* ii. 8. *Bolton, Fil. Brit.* 42, t. 23.

POLYPODIUM CARTHUSIANUM, *Villars, Hist. des Pl. Dauph.* iii. 842, according to Willdenow.

POLYPODIUM MULTIFLORUM, *Roth, Catalect. Bot.* i. 135.

POLYSTICHUM MULTIFLORUM, *Roth, Fl. Germ.* iii. 87.

POLYSTICHUM SPINULOSUM, *De Candolle, Fl. Franç.* ii. 561 (excl. syn. Swartz). *Ledebour, Fl. Ross.* iv. 515, in part.

POLYSTICHUM SPINULOSUM, β. DILATATUM, *Koch, Synops.* 2 ed. 979.

POLYSTICHUM DILATATUM, *De Candolle, Fl. Franç.* v. 241.

NEPHRODIUM CRISTATUM, *Michaux, Fl. Bor. Amer.* ii. 269; according to Pursh.

NEPHRODIUM DILATATUM, *Desvaux, Ann. de Soc. Linn. Paris,* 261.

DRYOPTERIS DILATATA, *A. Gray, Man. Bot. North. U. States,* 1 ed. 631.

LOPHODIUM MULTIFLORUM, *Newman, Phytol.* iv. 371; *Id.,* 1851, *App.* xvii.; *Id., Hist. Brit. Ferns,* 3 ed. 147 (excl. syn. Presl).

Var. **dumetorum**: fronds dwarf or dwarfish, oblong-ovate, or triangular-ovate, bipinnate, the stipites, rachides, and under surface of veins clothed with glands; pinnules convex, oblong, pinnatifid, sometimes wavy; sori large, with gland-fringed indusia; scales of the stipes broad-lanceolate, pale brown, whole-coloured or faintly two-coloured, fimbriate or jagged or glandular at the margin. [Plate XLVIII.]

LASTREA DILATATA, *v.* DUMETORUM, *Moore, Ferns of Gt. Brit. Nature Printed*, t. 25 (not of *Handb.* 2 ed. 124); *Id.*, *Handb. Brit. Ferns*, 3 ed. 125.
LASTREA DILATATA, *v.* MACULATA, *Moore, Handb. Brit. Ferns*, 2 ed. 124.
LASTREA DILATATA, *v.* COLLINA, *Moore, Handb. Brit. Ferns*, 2 ed. 123, in part.
LASTREA DUMETORUM, *Moore MS.* (Nat. Pr. Ferns).
LASTREA MULTIFLORA, *v.* COLLINA, *Newman, Hist. Brit. Ferns*, 2 ed. 222, in part.
LASTREA COLLINA, *Newman, Hist. Brit. Ferns*, 2 ed. 224, in part.
LASTREA MACULATA, *Deakin, Florigr. Brit.* iv. 110, fig. 1612.
ASPIDIUM DUMETORUM, *Smith, Eng. Fl.* iv. 281; according to the Smithian Herbarium.
LOPHODIUM COLLINUM, *Newman, Phytol.* 1851, *App.* xviii. in part; *Id.*, *Hist. Brit. Ferns*, 3 ed. 144, in part.

Var. **collina**: fronds narrow-elongate-ovate or ovate-lanceolate, bipinnate; pinnæ distant; pinnules convex, oblong, obtuse, the basal ones pinnatifid, the lobes very obtuse, serrated towards their apices with coarse acuminate teeth; scales of the stipes dark-centred, at the base numerous and subulately tipped, the upper ones few, broader. [Plate XLVII.]

LASTREA DILATATA, *β.* COLLINA, *Moore, Handb. Brit. Ferns*, 1 ed. 59; 2 ed. 123, in part; 3 ed. 125; *Id.*, *Ferns of Gt. Brit. Nature Printed*, t. 26 A—B. *Babington, Man. Brit. Bot.* 4 ed. 422.
LASTREA MULTIFLORA, *v.* COLLINA, *Newman, Hist. Brit. Ferns*, 2 ed. 222, in part. *Deakin, Florigr. Brit.* iv. 114, in part.
LASTREA COLLINA, *Newman, Hist. Brit. Ferns*, 2 ed. 224, in part.
LOPHODIUM COLLINUM, *Newman, Phytol.* 1851, *App.* xviii. in part; *Id.*, *Hist. Brit. Ferns*, 3 ed. 144, in part.

Var. **Chanteriæ**: fronds lanceolate, narrowed and truncate below, the apex caudately elongated, the stipites, rachides, and under surface glandular; pinnæ distant, the lowest unequally deltoid, the rest nearly equal; pinnules oblong, very obtuse, distant, pinnatifid, the lobes having coarse aristate teeth; indusia margined with small stalked glands; scales of the stipes lanceolate-aristate, entire, pale brown, with a dark central stripe. [Plate XLV.]

LASTREA DILATATA, *v.* CHANTERIÆ, *Moore, Ferns of Gt. Brit. Nature Printed*, t. 24; *Id.*, *Handb. Brit. Ferns*, 3 ed. 126.
LASTREA CHANTERIÆ, *Moore MS.* (Nat. Pr. Ferns).

Var. **angusta**: fronds linear-lanceolate bipinnate; pinnæ short deltoid, the anterior and posterior pinnules of the lower pinnæ very unequal; scales of the stipes narrow-lanceolate, pale-brown, two-coloured, pallid; indusia indistinctly glandular.

LASTREA DILATATA, *v.* ANGUSTA, *Moore, Handb. Brit. Ferns*, 2 ed. 124; 3 ed. 126; *Id., Ferns of Gt. Brit. Nature Printed*, under t. 22.

Var. **alpina**: fronds narrow, linear-lanceolate, membranaceous, bi-subtri-pinnate; pinnæ unequally deltoid; scales of the stipes broad-lanceolate, pale-brown, whole-coloured or variously two-coloured; sori large, with small fugacious ragged glandular indusia.

LASTREA DILATATA, *v.* ALPINA, *Moore, Ferns of Gt. Brit. Nature Printed*, under t. 22; *Id., Handb. Brit. Ferns*, 3 ed. 126.

Var. **nana**: fronds dwarf, ovate, bipinnate, somewhat glandular; pinnules decurrent, convex; scales of the stipes dark-centred; indusia small fugacious, the margin slightly glandular. [Plate XLVI.]

LASTREA DILATATA, *v.* NANA, *Moore, Handb. Brit. Ferns*, 2 ed. 127; 3 ed. 125; *Id., Ferns of Gt. Brit. Nature Printed*, t. 26 C—D.
LASTREA MULTIFLORA, *v.* NANA, *Newman, Hist. Brit. Ferns*, 2 ed. 222; 3 ed. 153. *Deakin, Florigr. Brit.* iv. 114.
LOPHODIUM MULTIFLORUM, *v.* NANUM, *Newman, Hist. Brit. Ferns*, 3 ed. 153.

Var. **tanacetifolia**: fronds ample, triangular or sub-triangular-ovate, tri-quadri-pinnate; scales of the stipes dark-centred; indusia small, the margin irregular, slightly glandular.

LASTREA DILATATA, *v.* TANACETIFOLIA, *Moore, Handb. Brit. Ferns*, 3 ed. 124; *Id., Ferns of Gt. Brit. Nature Printed*, under t. 22.
LASTREA DILATATA, *β.*, *Moore, Handb. Brit. Ferns*, ed. 1, 59.
LASTREA MULTIFLORA, *β.* DILATATA, *Deakin, Florigr. Brit.* iv. 113, 116, fig. *a.*
ASPIDIUM DILATATUM, *Willdenow, Sp. Plant.* v. 263. *Sprengel, Syst. Veg.* 106.
ASPIDIUM SPINULOSUM, *Schkuhr, Krypt. Gew.* 48, t. 47 (excl. *c.*)
ASPIDIUM SPINULOSUM, *β.*, *Hooker & Arnott, Brit. Fl.* 7 ed. 586 (excl. syn. Sm. and Newm.)
ASPIDIUM EROSUM, *Schkuhr, Krypt. Gew.* 46, t. 45 (monstrous).
POLYPODIUM TANACETIFOLIUM, *Hoffmann, Deutschl. Fl.* ii. 8.
POLYSTICHUM TANACETIFOLIUM, *De Candolle, Fl. Franç.* ii. 562; according to specimen from Prof. Fée.
POLYPODIUM ARISTATUM, *Villars, Hist. des Pl. Dauph.* iii. 844.

Var. **lepidota**: fronds short broadly-ovate, quadripinnate; ultimate pinnules small, distinct, pinnatifid and spiny-toothed; stipites and rachides everywhere densely clothed with lanceolate contorted whole-coloured scales. [Plate L.]

LASTREA DILATATA, *v.* LEPIDOTA, *Moore, Handb. Brit. Ferns*, 3 ed. 136.
LASTREA LEPIDOTA, *Moore MS.* (Handb. Brit. Ferns).

Var. **cristata**: fronds and pinnæ dilated or once or twice irregularly forked or crested at the apex. [Plate XLIX A.]

LASTREA DILATATA, *v.* CRISTATA, *Moore, Sim's Cat.* 1859, 10.

Var. **glandulosa**: fronds ample, lanceolate-ovate, or oblong-lanceolate, tripinnate below, densely covered with stalked glands beneath, as well as on the stipites and rachides; scales of the stipes pale whole-coloured, or faintly two-coloured, broadly lanceolate-ovate, semi-appressed. [Plate XLIV.]

LASTREA DILATATA, *v.* GLANDULOSA, *Moore, Handb. Brit. Ferns*, 2 ed. 124; 3 ed. 127; *Id. Ferns of Gt. Brit. Nature Printed*, t. 23. *Babington, Man. Brit. Bot.* 4 ed. 422 (excl. syn. Deakin).
LASTREA GLANDULOSA, *Newman, Phytol.* iv. 258.
LOPHODIUM GLANDULOSUM, *Newman, Phytol*, 1851, *App.* xviii.; *Id., Hist. Brit. Ferns*, 3 ed. 154.
LOPHODIUM GLANDULIFERUM, *Newman, Phytol.* iv. 371.

Caudex stout, usually erect, rarely decumbent, not creeping, often becoming elongated and trunk-like, sometimes tufted, the crown densely scaly; the fronds arranged in a circlet around the crown when the caudex is erect. *Scales* lanceolate-subulate, hair-pointed, brown, with a dark centre and paler margins. *Fibres* dark-brown, numerous, coarse, branched, tomentose.

Vernation circinate, the rachis often folded laterally as well as involutely fore and aft, the apex, however, being simply circinate.

Stipes terminal, and adherent to the caudex, variable in length, usually from about one-third to one-half of the entire length of the frond, stout at the base, green, densely scaly; the scales spreading, most numerous at the base, but usually abundant throughout the whole length of the stipes, and in the normal plant lanceolate-attenuate, and dark-centred like those of the crown, frequently almost black. *Rachis* convex behind, channelled in front, smooth, or in some plants otherwise normal, clothed with glands; somewhat scaly, especially at the back, with small subulate more or less distinctly two-coloured scales.

Fronds averaging two to three feet, but (exclusive of the varieties noticed below) varying from about a foot to five or six feet in length, and from six to sixteen inches in breadth, herbaceous, dark-green above, paler beneath, spreading, and more or less arched or drooping, ovate or ovate-lanceolate in the typical form, bipinnate or tripinnate. *Pinnæ* numerous, opposite or sub-opposite, the pairs more distant below. The lowest pair are obliquely-triangular elongate, the posterior pinnules being much larger than, often twice as large as, the anterior ones; the pinnæ of a few of the succeeding pairs have also an obliquely-deltoid outline, the obliquity gradually disappearing towards the upper part of the frond, so that those of about the third or fourth pair, as well as those above them, are nearly equal-sided: the upper pinnæ are also narrower, tapering very gradually from the base to the apex. *Pinnules* ovate-oblong, acutish, often convex, the basal ones stalked, the upper sessile and decurrent; the lower ones (especially those of the lowest pinnæ) are very deeply pinnatifid, sometimes pinnate, and the lobes or pinnulets are oblong and bluntish in outline. All the divisions are sharply-toothed, with teeth of subovate form, terminating in a bristle-like point or mucro, which is in general curved laterally towards the apex of the pinnule or lobe.

Venation in the pinnulets of the lower pinnæ, consisting of a stout flexuous *vein*, proceeding from the *costa* or rachis-like vein of the primary pinnule, forming a midvein, from which a *venule* proceeds into each marginal lobe, and this is forked where the lobe is toothed, so as to give off a branch towards each tooth, the anterior branch being fertile at some distance below its apex. In the larger of the less divided primary pinnules, the same arrangement occurs on a reduced scale, the *costa* producing a *vein* for each lobe, and this again a *venule* for each tooth, the lowest anterior venule only being fertile. The same arrangement, still more simplified, occurs in the smaller primary pinnules. The venules all terminate in a small club-shaped apex, below the tooth towards which they are directed.

Fructification on the back of the frond, and occupying the whole under surface. *Sori* numerous, variable in size, distinct, round, indusiate; medial sub-terminal or terminal, seated on the anterior basal venules in the less divided pinnules, and on the lowest anterior branch of the venules in the more compound pinnules; in the former consequently ranging in two lines, one on each side the mid-vein, and much nearer to it than the margin; in the latter forming two lines in a similar way along the lobes. *Indusium* reniform, rather large, convex, membranaceous, fringed around the margin with stalked glands; or sometimes smaller, flattish, and indistinctly glandular. *Spore-cases* numerous, brown, roundish-obovate. *Spores* roundish or oblong, muriculate.

Duration. The caudex is perennial. The fronds are semi-persistent, and under shelter endure throughout the winter though decaying at the base of the stipes. The young fronds are produced in spring, and additional ones uncertainly during the summer.

This is a most variable species, extremely difficult to understand, in consequence of its polymorphous character. It is more or less intimately united with two or three allied species, by means of transition forms, the kindred British plants being *Lastrea æmula*, on the one hand, and that known as *Lastrea spinulosa*, on the other. The latter is distinguished readily enough by its creeping caudex, by the few broad pallid scales of its stipes, and by its entire indusium; the former by its more strictly evergreen habit, by its lacerated scales, its anthoxanthoid fragrance, and by the absence of stalked glands from the margin of its indusium. *Lastrea æmula* may also be known by the concavity of its pinnæ and pinnules; and even in the decay of its fronds it is peculiar, for whilst the forms of *cristata* and *dilatata* decay first near the base of the stipes, so that the fronds often fall while they yet appear green and fresh upwards, in *æmula* the stipes continues firm, while the frond itself is undergoing decay, the disorganisation going on from above downwards, and not from below upwards. The distinguishing marks of *Lastrea dilatata*, in the group of which its variations form so large a proportion, are, its lanceolate dark-centred scales, and its gland-fringed indusia.

The typical form of this protean species is generally distributed over the United Kingdom, occurring from the coast level, to an elevation of 3600 feet or upwards. It prefers shady situations, such as moist woods and glens, thickets, and sheltered hedge banks. The variety *tanacetifolia* is also abundant, and probably universally distributed throughout Great Britain and Ireland. The other varieties are rare, or local. The species itself has been recorded from the following stations:—

Peninsula.—Cornwall: Penryn, *G. Dawson.* Devonshire: Lynmouth; Torquay; Walkhampton; Hartland, *Rev. J. M. Chanter;* Barnstaple, *F. Mules;* Linton; Ilfracombe (various forms), *Rev. J. M. Chanter;* valley of the Erme, *H. F. Dempster,* etc. Somersetshire: Inglishcombe Wood, *R. Withers;* Selworthy.

Channel.—Hampshire. Isle of Wight: Newnham, near Ryde. Dorsetshire. Wiltshire: Spye Park; Alderbury Common, Salisbury. Sussex: Tilgate Forest; Hastings; Tunbridge Wells; Eridge Rocks.

Thames.—Hertfordshire: Hertford Heath; Broxbourne; Aldenham; Hitchin, etc. Kent: Chislehurst (with a glandular form). Surrey: Barnes; Chertsey; Bagshot; Virginia Water, and other parts. Middlesex: Hampstead (with a glandular form). Essex: Epping. Oxfordshire; Chipping Norton, *H. Buckley.*

Ouse.—Norfolk: Norwich; between Lynn and Bawsey, *Dr. Allchin.* Cambridgeshire. Northamptonshire.

Severn.—Warwickshire: Stoke Heath; Stinchall; Whitley, and other parts. Gloucestershire: Forest of Dean. Monmouthshire: Pen-y-garn and Trevddun, *T. H. Thomas.* Herefordshire: Howle Hill, Ross; Colwall. Worcestershire. Staffordshire: Wolverhampton, *Mrs. Rutter;* near Stafford, *Rev. R. C. Douglas.* Shropshire: Titterstone Clee (with a glandular form); Sandford Heath; Hawkestone; Bomere; Sutton Spa, and Shomere, near Shrewsbury; Pimhill; Shawbury Heath, *Rev. W. A. Leighton;* Whitchurch, *R. W. Rawson.*

Trent.—Leicestershire. Derbyshire: Cromford Moor, near Matlock. Lincolnshire. Nottinghamshire.

Mersey.—Cheshire: Lindon Moss, near Mobberley; Ashton Moss, *H. Buckley.* Lancashire: Risley Moss, near Warrington; Clough, near Manchester; and elsewhere on the hills.

Humber.—Yorkshire: Leckby Car; Heckfell Wood; Sheffield Moor; Bradford, *J. T. Newboult;* Settle; Halifax; Thirsk; Ingleborough, and elsewhere.

Tyne.—Northumberland: Morpeth. Durham: Sunderland.

Lakes.—Westmoreland: Elter Water, *Rev. G. Pinder;* Langdale; Silverthwaite; Old Man; and the Rocky Fells throughout the district. Cumberland: Red House. N. Lancashire: Torver, near Conistone. Isle of Man.

S. Wales.—Brecknockshire: Brecknock Beacon; Drygarn, *J. R. Cobb.* Radnorshire: Gamrhin, above Rhayader, *J. R. Cobb.* Glamorganshire. Cardiganshire. Pembrokeshire: Castle Malgwyn, *W. Hutchison.*

N. Wales.—Anglesea: Cickle, *Rev. W. A. Leighton.* Denbighshire: Ruthin, *T. Pritchard;* Wood at Glan-r-avon, *T. P.;* Rhuabon, *A. L. Taylor.* Flintshire. Merionethshire: Festiniog. Carnarvonshire: Aber, *Rev. W. A. Leighton.*

W. Lowlands.—Dumfriesshire. Ayrshire: Maybole, *W. Dickson.* Lanarkshire.

E. Lowlands.—Roxburghshire. Berwickshire. Edinburghshire: Habbie's How, *E. Hall;* Hawthornden, *T. M.*

E. Highlands.—Stirlingshire. Clackmannanshire. Kinross-shire. Fifeshire: Lomond Hills. Perthshire: Dunkeld; Ben Lawers; mountains near Crieff; Pass of Trosachs; Loch Katrine; etc. Forfarshire: Ingelmady; Dundee; Den of Fullerton, *A. Croall.* Kincardineshire: Kingcausie, *J. T. Syme.* Aberdeenshire: Benna-Baird; Glen Callater, Braemar, *A. Croall.* Banffshire. Morayshire. E. Inverness-shire.

W. Highlands.—Argyleshire: Appin, *J. T. Syme;* Poltalloch; near Loch Ballenoch; Ardrishiag, *T. M.;* Cairndow; Glen Croe. Dumbartonshire: Tarbet, *T. M.* W. Inverness-shire. Isles of Arran, Islay and Cantyre. Ailsa Craig.

N. Highlands.—Ross-shire. Sutherlandshire. Caithness, *T. Anderson.*

N. Isles.—Hoy, and other islands of Orkney, *T. Anderson.*

W. Isles.—N. Uist. Harris. Lewis.

Ulster.—Down: Newtown, *Dr. Mateer.*

Connaught.—Galway: Connemara; Kylemore, and Glendalough, *R. Barrington.*

Leinster.—Wicklow: Newtown, *Miss Tarbet;* Powerscourt waterfall; valley of Glencullen, *R. Barrington*; near Upper Lough Bray, *R. B.* Dublin Mountains, *J. R. Kinahan;* top of Three Rock Mountain, 1400 ft., *R. Barrington;* Glen near Kingston, *R. B.* Kilkenny. King's.

Munster.—Waterford; near Clonmel, *J. Sibbald.* Clare. Limerick. Tipperary.

Channel Isles.—Jersey. Guernsey, *C. Jackson.*

Lastrea dilatata is a common and generally dispersed European species, occurring from Lapland and Norway to Portugal and Spain, the Savoy Alps, Italy, Croatia and Transylvania. In Asia it occurs in Kamtschatka, near Petropaulowski, and in Mingrelia; in Africa, in the Azores and in Bourbon. In America it occurs at Sitka, and at Kodiak in the Russian territory; at Port Mulgrave, and in the Rocky Mountains; in New England and Canada. There is also in the Hookerian Herbarium a specimen labelled from New Zealand.

This is a plant of easy culture. In almost any kind of soil that can be kept moderately moist, and in any situation where it will enjoy a moderate amount of shade, it will grow readily; and it is really a very handsome plant for rockeries, always moreover easily obtainable. Few indeed if any of our large growing species are more elegant, its broad and compound fronds being of a rich deep green and enduring; they are in fact subevergreen under shelter. The best soil for it and its several varieties, is a mixture of turfy peat and loam with sand; and they all succeed very well cultivated in pots.

This Fern is very prolific of varieties. The most remarkable among them, and those which we consider as being of botanical importance, have been already briefly mentioned, but we propose in this place to give more complete descriptions of them, and to notice their distribution, as well as to enumerate such other forms as seem to call for mention in accordance with the plan we have followed in other cases. We may, however, observe that while the modifications of form assumed by this polymorphous species are numerous, the varieties have in some cases less definite characters and limitations than in the species we have previously dealt with. Some of the forms, on the other hand, are very marked and permanent.

1. *pumila* (M.). A dwarf subdeltoid or ovate-deltoid bipinnate form, in which the scales are for the most part pallid but two-coloured. It is the plant called *dumetorum* in the earlier editions of our *Handbook of British Ferns*, where, as we now believe, it was mistaken for Sir J. E. Smith's plant. It may perhaps be the young fertile condition of some of the larger-growing forms, but we incline to believe that there exists a dwarf permanent state, such as that we here describe. In some specimens quite similar in appearance to those with the pallid scales, the scales are more strongly two-coloured, but the pallid-scaled plants are more frequent. Specimens which we refer here have been obtained from—Middlesex: Hampstead. Worcestershire: Malvern. Yorkshire: Wentworth. Devonshire: Ilfracombe, *Rev. J. M. Chanter;* Ham, near Plymouth, *Rev. C. Trelawny.* Pembrokeshire: Castle Malgwyn, *W. Hutchison.* Denbighshire: Rhuabon, *A. L. Taylor.* Carnarvonshire: Aber, *Rev. W. A. Leighton.* Perthshire: Corrach Uachdar. Dumbartonshire: Tarbet. Forfarshire: Guthrie Woods, *A. Croall.* Aberdeenshire: Glen Callater, Braemar, *A. Croall.* Arran. Bute: Rothesay. Galway: near Clifden, Connemara, *R. Barrington.* Donegal: Killybegs, *R. B.* Wicklow: The Dingle, *R. B.;* Glen of the Downs, *R. B.* Dublin Mountains, *J. R. Kinahan.*

2. *dumetorum* (M.). This is a very distinct variety, in some at least of its forms. The most marked, which we take as the type of the variety, is one found in the Lake district by Miss M. Beever, a dwarf or dwarfish form, with broad-ovate, or elongate-triangular, or sometimes deltoid fronds, remarkable for their glandular surface, and their large abundant sori produced freely on plants of very immature age. This form, or variety, which appears to us even to have some title to specific distinction, occurs under several modifications, some of which have been referred to the var. *collina,* from which, however, they differ in their abundant glands, and fimbriated or jagged scales. Our typical form, that found in Westmoreland

by Miss M. Beever, to whom we are indebted for specimens, sufficiently accords with the imperfect examples of the *Aspidium dumetorum* in Sir J. E. Smith's herbarium, as to be considered the same plant, and we so regard it. This plant has elongate-triangular-ovate fronds seldom exceeding a foot in height, and very glandular, especially on the stipites, rachides, and lower surface of the veins; they are bipinnate, the pinnæ concave and bluntish; the pinnules broad oblong, or oblong-ovate, convex, crispy, and coarsely-toothed, the teeth broad and acuminately tipped by a small bristle. The stipes is sparingly clothed with lanceolate scales of variable width, and of a pale-brown colour, sometimes scarcely at all darker in the centre, sometimes distinctly two-coloured, but always having their margin fimbriate. The sori are large, distinct, produced over the whole under surface, and covered by indusia, which are prominently fringed with stalked glands. Young plants of this form, but a few months old, and three or four inches high, bear fronds which are abundantly fertile. Miss Beever's plant was gathered in Silverthwaite, and in other rocky Fells of that part of Westmoreland; and the same form has been obtained by Mr. F. Clowes from Hawes Water, and by the Rev. G. Pinder, from Elter Water. Sir J. E. Smith's *dumetorum* came from Cromford Moor near Matlock, in Derbyshire; from Westmoreland; from Mount Glyder overhanging Llyn Ogwen in North Wales; from near Phainon Vellon; and from Rivelston Wood, near Edinburgh. The plants from the following habitats we believe to be the same variety, inasmuch as they agree in the pale-coloured broad lance-shaped, fimbriated scales of the stipes and crown, in the dwarf habit, and subtriangular or ovate fronds, in the glandulose surface, and the large distinct sori; and differ from that first described, in little but the absence of the crispy aspect of the pinnules, while they present but trifling variation among themselves:—Devonshire: Challacombe, *Rev. J. M. Chanter;* Ilfracombe, *J. Dodds.* Lancashire, *R. Morris.* Cumberland, *Mrs. Delves.* Isle of Man, *Dr. Allchin.* Yorkshire: Ingleborough, *T. Blezard.* Radnorshire: Gamrhin, above Rhayader, *J. R. Cobb.* Carnarvonshire: Snowdon, *Mrs. Jennings;* Moel Siabod. Argyleshire: Loch Eck, near Dunoon, *G. R. Alexander;* Ardrishiag, *Miss F. Griffith;* near Loch Ballenoch. Dumbartonshire: Tarbet, *T. M.*

Arran: Goat Fell. Aberdeenshire: Braemar, *A. Tait.* Ross-shire: Dingwall (sori smaller), *Miss Murray.* Caithness: Thurso, *A. Tait.* Wicklow: Glen of the Downs, *R. Barrington.* Kerry: Killarney, *R. B.* The Arran, Devon, and Isle of Man forms have their scales somewhat less fimbriated than the others. A small ovate form, agreeing with the foregoing in the scales and in the glands, found in Glen Croe, Argyleshire, is tripinnate at the base, and has the pinnules much smaller than usual.

3. *collina* (M.). This is another distinct and permanent variety, approaching *dumetorum* (2) in some respects, but obviously different in others. It is a remarkably neat-looking and elegant plant, of erect habit, having sometimes an ovate outline of frond, attenuately elongated at the apex, but also occurring of a more elongated, *i. e.*, an oblong-lanceolate or ovate-lanceolate outline. The fronds are dark green, one to two feet high, smooth, or sparingly glandular, bipinnate. The stipes is variable in length, both in the wild specimens, of which some are found beneath masses of rock, and also under cultivation; it is from one-half to one-third the length of the fronds, green above, tinged with dark purple-brown at the base, scaly, with entire lanceolate dark-brown scales, which have a conspicuous darker central mark. The scales at the base of the stipes, where they are most numerous, are narrow, and have a long subulate point; higher up they are scattered, and many of them broader and shorter; and the rachis itself is almost devoid of scales. The pinnæ, especially the lower ones, are distant and spreading: the lowest pair unequally deltoid; the next pair more elongate and less unequal; and the remainder narrower, parallel-sided, rounding slightly near the end to an acutish, not at all acuminate, point. The pinnules are convex, obtusely oblong-ovate, the basal ones narrowed to a broadish stalk-like attachment, the rest sessile and more or less decurrent; the larger pinnules are deeply pinnatifid with blunt oblong lobes, which are sparingly toothed, the teeth coarse acuminately aristate, occurring mostly at the apex. The sori are for the most part arranged in two lines along the pinnules, as in the smaller forms of the species, and are covered by gland-fringed indusia. This variety was first brought into notice by the Rev. G. Pinder, to whom we are indebted for specimens, and it is from his

plant found at Elter Water, in Westmoreland, that the foregoing description has been drawn up. Miss S. Beever has communicated the same variety from Torver, near Conistone, where it was collected by Mr. T. Ecclestone; this latter plant is rather larger and more divided, and has prettily concave pinnæ, and strongly convex pinnules, and the plant is also somewhat glandular, which is hardly, if at all, the case with that we have taken as the type of the variety. Mr. Pinder found the latter extending over the Lake district of Westmoreland and Lancashire, and on the Yorkshire hills. The stations recorded are—Devonshire: Ilfracombe. Yorkshire: Ingleborough. Westmoreland: Langdale; Mardale; Hawes Water, *F. Clowes.* ? Carnarvonshire: Tre'r Ceiri, *C. C. Babington.* Forfarshire: Guthrie Woods (larger and rather more divided), *A. Croall.* Dumbartonshire: Tarbet. Argyleshire: Ardrishiag. Arran, *T. M.* ? Wicklow: Powerscourt Waterfall, *C. C. Babington.*

4. *Smithii* (M.). This is a small plant, and has something of the general aspect of *collina* (3), but the fronds are more oblong, and the pinnæ more equal-sided. The fronds are about a foot high, with a stipes of three inches; the pinnæ of the lower half of equal length, and with the tapering apex giving a narrow elongately subtriangular-ovate outline. The pinnæ are opposite, horizontal, distinct, and having but slight inequality, even in the lowest, in the size of the anterior and posterior pinnules. The pinnules are set on at a right angle, the basal ones with a narrow attachment, the rest narrowly decurrent on the rachis, ovate-oblong, obtuse, the basal ones pinnatifid, the lobes blunt, with distinct acuminate teeth. The scales of the stipes are dark, two-coloured, lanceolate, narrower and more elongated about the base of the stipes. The plant is doubtless related to *collina*, and is, perhaps, only a modification of it. The Irish forms of this affinity are little known, and require a more complete investigation than has hitherto been given to them; and the same remark applies to the Irish forms related to *dumetorum.* We describe *Smithii* from a frond sent by Mr. H. Shepherd, of Liverpool, who states that the plant was given to him by Dr. Mackay, as that from which Sir J. E. Smith drew up his description of *Aspidium spinulosum;* so that it is probably the plant from Spike Island, near Cove, below Cork, mentioned in *English Flora* (iv. 279).

5. *obtusa* (M.). This form does not well associate with any of the others that we have seen. It is of medium size; the fronds narrow ovate in outline, with oblong obtuse shallow-lobed pinnules, set on nearly at a right angle, and having few coarse acuminate teeth. We have found this form well marked at Hampstead, Middlesex; others collected at Hastings, Sussex, resemble it.

6. *distans* (M.). This is a marked variety, somewhat resembling *Chanteriæ* (7), but it is smoother, more lax in habit, and apparently growing to a larger size. The outline of the frond, which grows three feet high or more, is ovate. The pinnæ are distant, and scarcely enlarged on the posterior side in the upper parts of the frond, the lowest pair only being very oblique, and the next pair slightly so. The pinnules are set wide apart on the rachis, and are ovate-oblong, obtuse, the narrowed stalk-like base somewhat decurrent, except in the very lowest pinnules; the basal pinnules are deeply, the rest shallowly divided into short oblong obtuse lobes, which are coarsely toothed, the teeth acuminately-aristate. The sori are numerous, forming two lines near the midrib; the indusium is slightly glandular. It was found at Coombe Wood, Surrey, by Mr. S. F. Gray.

7. *Chanteriæ* (M.). This is an elegant and a remarkably distinct form of the species, differing obviously in the narrowed form and attenuated apex of its fronds, its distant pinnæ, and its distinct blunt pinnules. The stipites, rachides, and under surface of the fronds are clothed with sessile or very shortly-stalked glands. The stipes bears numerous lanceolate and ovate-lanceolate entire scales, which are of various sizes, brown, with a dark central streak, and tipped by a longish, weak, bristle point. The fronds grow about a couple of feet in height, and are nearly erect in habit, and lanceolate or oblong-lanceolate in form, the base narrowing yet terminating abruptly, and the apex attenuated and caudate; those of the cultivated plant, being more lax, are often ovate with an elongated point; they are always bipinnate, sometimes almost tripinnate. The pinnæ are distant, somewhat spreading, and more or less twisted, so that the upper surface is directed towards the zenith; the lowermost pair, about three and a half inches long, and an inch and a half broad, are very unequally deltoid, their posterior basal pinnules being

more than twice the length of the anterior ones, and these posterior pinnules are themselves almost pinnate; the next pair is unequally deltoid, but the posterior pinnule is only about one-third longer than the anterior; and the inequality is nearly lost in the next and the succeeding pinnæ, which narrow gradually to the apex, the longest about the centre of the frond being about five inches long and an inch broad. The basal pinnules of the upper pinnæ are nearly oblong, their base being but little broader than their apex, which is very blunt; and they have a narrow stalk-like attachment, which becomes broader and more decurrent in the pinnules higher up the pinnæ. The pinnules of the lower pinnæ are more or less deeply pinnatifid according to their position, and the lobes, which are bluntly oblong, have a few coarse distinct teeth, each of which is terminated by a bristle-like point. The sori are small, numerous, forming a line along each side near the midrib of the smaller pinnules, and along the lobes of the larger ones; they are covered by reniform indusia, which are fringed with small stalked glands at the margin. This very marked variety, which we have found to be quite constant, and renewable from the spores, was discovered in 1854, by the Rev. J. M. Chanter and Mrs. Chanter, after whom it has been named, at Hartland, on the north coast of Devon, where it was met with growing in moderate quantity within a limited area, and accompanied and surrounded by other common forms of the species. A similar plant, which may perhaps prove the same, has been gathered at Challacombe, Exmoor, by Mr. H. F. Dempster.

8. *angusta* (M.). This variety has the outline and general features of the erect typical form of *spinulosa* (*L. cristata*, var.), but it possesses also the particular characteristics of *dilatata*. The fronds are narrow linear-lanceolate, about two feet high, bipinnate; the stipes being as long as the lamina, and furnished rather scantily with large lance-shaped pale-brown dark-centred scales. The pinnæ are shortly deltoid, and the lower two or three pairs very unequally so, the posterior pinnules being much the largest. The pinnules are narrow, oblong, obtuse, deeply pinnatifid with ovate or oblong lobes, having aristate teeth. The sori are small, abundant, occurring from the base to the apex of the frond, and covered by small indistinctly glandular convex indusia. The variety was established in our

Handbook of British Ferns (2 ed. 124), on two fronds gathered by the late Miss Bower, near Tunbridge Wells. We have subsequently seen a somewhat similar form, scarcely identical, from Hartland, Devon, where it was found by Mrs. Chanter in company with *Chanteriæ*.

9. *alpina* (M.). This is an elegant form, remarkable for its delicate texture, much more delicate and membranaceous than in any other form of the species we have seen. The fronds seem to be normally oblong, that is, nearly straight-sided with the point tapered off, such as occurs in the typical state of *spinulosa;* but some of our specimens are ovate, or even broadly ovate, probably resulting from differences of age, or of the conditions under which they were grown. The fronds are almost or quite tripinnate below, bipinnate upwards. The pinnæ are ascending, membranaceous in texture, obliquely deltoid or ovate below, ovate-lanceolate, and nearly equal above; the lowest pinnæ are very unequal sided, and in the oblong, or as we regard them typical fronds, very little shorter than several of the succeeding pairs; but the rest, above the second pair, are very slightly unequal. The pinnules are rather ovate, or elongate-ovate, according to their position, the lowermost ones almost or quite cut up into ovate-oblong pinnulets, which are lobed, the lobes serrate; the smaller ones are deeply pinnatifid with mucronate acute serratures. The sori are large numerous, placed near the base of the sinuses, and so forming generally two lines along the pinnules; they are furnished with small fugacious indusia having a ragged somewhat glandular margin. The scales are broad lance-shaped entire, sometimes whole-coloured palish-brown, sometimes, and apparently most commonly, pale brown with a dark central mark varying in intensity. This form occurs plentifully among rocks on the higher parts of Ben Lawers, Perthshire, where it first attracted our notice. It occurs also in—Aberdeenshire: Glen Callater, Braemar, *A. Croall;* Loch-na-gar (a dwarfed and depauperated form), *W. Sutherland.* Westmoreland: Hawes Water, *F. Clowes.* Yorkshire: Ingleborough, plentiful on the N. E. side, *T. Blezard.* Lancashire: near Shooter's-Spring, Salter-fell, Roeburndale, near Lancaster, *T. B.* A very dwarf form with ovate fronds, found by Dr. Balfour on Ben Voirlich in Perthshire (as *montana*), and which does not well associate

with any other form, is probably a small state of the var. *alpina;* as also, a similar dwarf form, found on Loch-na-gar in Aberdeenshire by Mr. W. Sutherland.

10. *nana* (Newm.). This form proves to be a permanent variety, and not an immature condition of the species, as might be supposed. It differs most obviously from the usual and commoner forms of the species in its constantly smaller size; the extreme length of the fronds, including the stipites, varying from two to four inches in the smallest forms, to eight or ten inches or a foot in the largest forms of the variety. This small size and dwarfness is a permanent characteristic, the variety having been observed for the last twenty years, by Mr. J. Tatham, growing near Settle, in Yorkshire, without change, and in company with the ordinary forms of the species three feet in height; and even when freely manured, these plants though attaining about fifteen inches high, did not lose the dwarfish aspect of the natural specimens. The Rev. J. M. Chanter has also observed the same fact of constancy for a series of years in plants which we think belong to this variety, occurring near Ilfracombe, in Devonshire; and cultivation in a greenhouse was not found to add to the size of the Devonshire plants, which assume slight variations of form among themselves. The Settle plant is the typical form of the variety. The fronds of this are ovate broadest at the base, or oblong-ovate, bipinnate. The stipites and rachides, as well as the under side of the veins, are sparingly clothed with short-stalked glands; and the stipes moreover, which forms nearly half the entire height, is clothed thickly at the base, more sparingly upwards, with lanceolate scales having the usual dark central mark. The pinnæ are spreading and somewhat acuminate, the lowest pair unequal-sided, but the rest nearly equal. The basal pinnules are distinctly stalked, the next decurrently stalked, and the upper ones adnate, somewhat convex, the larger ones deeply, the rest shallowly lobed, the lobes being serrated; the smaller ones are merely serrate; the teeth all acute and mucronate. The sori are often most copious in the upper part though frequently occupying the whole of the frond, and form a line on each side the midvein of the pinnules nearer the rib than the margin; they are rather small, and are each covered by a delicate, somewhat glandular-margined indusium, which soon shrivels and

becomes concealed among the spore-cases. The plants from Ilfracombe are very similar, but consist of two forms differing slightly in the colour of the scales, and in the form and manner of the toothing of the pinnules; the dwarfer plant having the more pallid scales. Besides the Settle and Ilfracombe plants, there are others which we refer to this variety, from—Devonshire: Challacombe, Exmoor. Dumbartonshire: Tarbet. Mayo: foot of Slieve More, near Dugort, Island of Achill, *R. Barrington.* Wicklow: Glen of the Downs, *R. B.* Kerry: Killarney, *R. B.* It is probably not uncommon in elevated rocky localities; according to Mr. Newman it is frequent in the hill districts of Scotland, Wales, and Ireland.

11. *micromera* (M.). The peculiarity of this form, which has a stout stipes clothed with large very dark scales, and is of the normal ovate-lanceolate outline, and about two feet high, is, that it is more finely divided than usual. The fronds are not large, but they are almost quadripinnate, and the lobes are small and have numerous small sharp teeth. It was found in the neighbourhood of Ilfracombe, Devonshire, by the Rev. J. M. Chanter. Somewhat similar forms occur in—Devonshire: Barnstaple, *C. Jackson.* Pembrokeshire: Castle Malgwyn, *W. Hutchison.* Argyleshire: Glen Croe, *T. M.*

12. *deltoidea* (M.). This form grows about two feet high, and has deltoid tripinnate finely-cut fronds, the stipes slender, and the whole aspect of the plant light and elegant. The scales of the stipes are dark-coloured. The pinnules and lobes are rather blunt, with largish mucronate unequal teeth. It is a Devonshire plant, collected by the Rev. J. M. Chanter; and a very similar one has been gathered at Barnstaple by Mr. C. Jackson.

13. *fuscipes* (M.). This is a glandular form of elegant appearance, growing two feet or more in height, and having the stipites, which are comparatively slender, of a pale chestnut brown behind, and furnished with dark-coloured or sometimes palish narrow scales. The fronds are broad ovate, almost triangular, rather delicate in texture, glandular, tripinnate below, the points of the frond and of the pinnæ acuminate or sometimes almost caudate. The segments are oblong, the largest lobate and serrate, the smaller merely serrate; the teeth are everywhere large and mucronate, and occur towards the upper end of the segments, the bases of which are rather narrowed

and entire. The form to which the name was first given was collected in Guernsey, by Mr. G. Wolsey. Some very similar forms are less decidedly tripinnate and others less glandular. We associate with the typal Guernsey plant others from—? Devonshire: Torquay (not glandular), *J. Carton.* Pembrokeshire: Castle Malgwyn, *W. Hutchison.* Denbighshire: Ruthin, *T. Pritchard.* Dublin: Glendruid, *R. Barrington;* Three-rock mountain, *R. B.* Kerry: Killarney, *R. B.* Donegal: Gweedore, *R. B.*

14. *tenera* (M.). This is a very fine and elegant glandular form, nearly related to *fuscipes* (13), but more delicate in texture, and more divided. The stipes is pale chestnut brown behind, furnished towards the base with dark-coloured lanceolate scales, and abundantly glandular upwards, as also are the rachides and the veins of the under surface of the frond. The texture is extremely thin and delicate. The fronds are two to three feet high, ovate, tripinnate. The pinnæ are broad and caudately acuminate. The posterior pinnules of the two lower pairs of pinnæ are elongate oblong-acuminate, the rest shorter and obliquely ovate, but also acuminate. The pinnulets are sessile below, adnate upwards, the basal ones generally obliquely-ovate, but the greater number are narrower, often somewhat falcate, and with a tendency to the development of an auricle-like basal anterior lobe; they are deeply-lobed or serrated, with acute mucronate not very prominent teeth. It has been sent to us from Windermere, by Mr. F. Clowes, and though similar in some respects to *fuscipes*, appears to be a distinct form, its texture, division, and manner of toothing furnishing its distinctive marks.

15. *valida* (M.). This form is stout, erect, rigid, fleshy-looking, becoming thick and leathery when dry. The fronds are large and broad, ovate in outline, bipinnate, or more frequently tripinnate. The stipes is stout, furnished not very abundantly with lanceolate scales, which are two-coloured but variable in intensity. The pinnæ are broad and rather crowded. The pinnules are divided almost to the midrib over the greater part of the fronds, and when these are large, quite so, in the case of the basal pinnules; they are oblong ovate, a little curved forwards, the lobes oblong obtuse, lobate-serrate, with bristle-tipped teeth. The venules each terminate near the

margin on the upper surface in a hair-like white line, which gives the plant a falsely strigose appearance. It has been sent to us from —Devonshire, *Rev. J. M. Chanter.* Somersetshire: Nettlecombe, *C. Elworthy.* Sussex: near Tunbridge Wells, *W. W. Reeves.* Guernsey, *C. Jackson.*

Mr. Tait has sent us a somewhat similar form from Monkland Glen, near Airdrie, Lanarkshire. This form (*erecta*) has long stipites, and broad ovate almost triangular fronds, which grow very erect, in the way of *spinulosa;* the pinnæ are distinctly concave, while the pinnules are convex, producing a crispy appearance; it grows two feet or more in height.

16. *tanacetifolia* (M.). This is a common tripinnate state of the species, with broad ovate fronds, having a strong tendency towards a triangular outline, which is even sometimes acquired. The fronds are usually large, though plants of but moderate size occur, in which the peculiarities of the variety are fully developed. The stipes has entire lanceolate dark-brown abundant scales, marked with a still darker bar down their centre, as in all the common forms. It is a variable form, merging insensibly into that which we have considered as the type of the species, but when large and lax is very handsome. We are indebted to Professor Fée of Strasburg for a specimen of the *Polystichum tanacetifolium* of De Candolle, which has enabled us to identify it with this form of *Lastrea dilatata.* It seems to be common. We have received specimens from—Devonshire: Hartland, *Rev. J. M. Chanter.* Kent: Tunbridge Wells, *Miss Bower.* Surrey: Chertsey. Middlesex: Hampstead. Hampshire: Breamore. Worcestershire: Daylesford. Yorkshire: Heckfell Woods. Argyleshire: Glen Gilp, Ardrishiag. Dumbartonshire: Tarbet. Aberdeenshire: Glen Callater, Braemar, *A. Croall.* Mayo: Westport, *R. Barrington.* Sligo: Lough Gill, *R. B.* Down: Balliivy, *R. B.;* Mourne Mountains, *Hb. Macreight.* Wicklow: Glen of the Downs, *R. Barrington.*

17. *lepidota* (M.). This is a most remarkable variety, so thoroughly distinct in character that we believe it is entitled to specific rank; and we notice it in this subordinate position only because the exact evidence of its being a native plant is wanting. There is, however, little or no doubt of its British origin. Though a comparatively

dwarf plant, it is conspicuously more divided than any other British *Lastrea*. Its chief peculiarities consist in the fronds being quadripinnate, the pinnules small and distinct, and the stipites and rachides everywhere densely lepidote-scaly. The fronds are about a foot and a half high, very broadly ovate: indeed the lamina is almost as broad as long, quadripinnate on the posterior side of the lowest pinnæ. The pinnæ are very unequal throughout the frond, the posterior ones being much the largest. The primary pinnules are distant on the rachides of the pinnæ, elongate ovate in outline, everywhere pinnately divided, the lowest posterior basal ones also unequal-sided. The secondary pinnules or pinnulets are distant, short ovate, and bluntish, with a tapered stalk-like base; below they are divided into broad pinnule-like lobes (which become tertiary pinnules on the basal posterior pinnule of the lowest pinnæ), and at their apex they are cut into coarsish mucronate incurved teeth, the lobes themselves being also similarly toothed. These ultimate divisions are comparatively small. The main stipites and rachides are densely clothed with large brownish entire lanceolate or ovate scales, which become smaller and shorter (not narrowed and hair-like) upwards; and both the secondary and tertiary rachides are conspicuously clothed with these smaller shorter scales, so that the appearance is rather lepidote than setiferous; the scales are often contorted. In our plants, cultivated under glass, a peculiarity of development is manifested, the evolution of the frond being indefinite, so that the basal pinnules of a pinna become fully grown with mature ripened or scattered sori, before the point of the same pinna is unrolled. The central part of the frond has consequently shed all its spores, while its apex, and the apices of the pinnæ are still growing on. We have had no opportunity to observe if this habit is maintained in out-door culture. The plant was first noticed in the collection of Mr. Tait, of Edinburgh, who had obtained it from Mr. Stark, a nurseryman of that city, with the information that it had been procured from Yorkshire. No further light on its history can be obtained from this source, and it is to be hoped that our very characteristic plate will enable some one to certify its British origin. We know of no exotic fern at all like it. [Plate L.]

18. *decurrens* (M.). This is an abnormal dwarf form, of very

peculiar aspect. The fronds are of triangular outline, not more than a foot high, including the stipes of about three inches, which latter is sparingly furnished with the dark-centred lanceolate scales belonging to the species. The fronds are not all alike, but the triangular outline with here and there a defect is maintained. The usual obliquity of the basal pinnæ is sometimes but not always developed; and the pinnæ are here and there, but not often, wanting or much abbreviated. The pinnules are rather distant, narrow oblong, all pointing forwards, narrowed below, and more or less distinctly decurrent at the base; they are very distantly lobed, the basal lobes being generally short and depauperated, some of the more perfect upper ones having a few coarse spiny teeth. Some fronds are rather less decurrent, and the pinnules rather more deeply divided, the lobes being in this case more generally and equally developed, and all of them irregularly spine-toothed. In one frond before us, the apices of all the pinnæ are equally abortive, producing a narrow triangular frond with truncate pinnæ, and lobed as in the former. The most normal-looking fronds we have seen, are of the same small size, but with oblong pinnules, which are cuneate as well as decurrent at the base, the margin lobed, and the lobes irregularly mucronate-serrate; some of the pinnules, here and there, being abortive or depauperated. It is an interesting monstrosity, and seems constant. The root was gathered near Scarborough in Yorkshire, in 1855, by Mr. Clapham, and has since maintained the same character.

19. *minima* (M.). This is a pigmy fern, the fronds being from about three and a half to seven inches in total stature. The fronds have an ovate outline, more or less affected by the parts becoming depauperated. The more normal parts resemble *decurrens* (18), only they are broader and more perfect, the pinnules in these portions being oblong, lobate-serrate, with acuminate teeth, often narrowed below and generally decurrent. The pinnæ near the base of the fronds are many of them depauperated; some reduced to the size of pinnules, with incised edges; some much shortened, with the pinnules also reduced, and often consisting only of a few coarse lobe-like teeth. The pinnæ are either wholly affected in this way, or some of the pinnules among the more perfect ones acquire this

pigmy character, so that the fronds are very diverse. The plant is constant to these features, and is an ornamental dwarf form of the species. It was found in Cant Clough, near Todmorden, Lancashire, by Mr. A. Stansfield, who describes it as very dwarf, constant, and beautiful.

20. *pygmæa* (M.). Another pigmy form, found by Mr. Stansfield in the same locality with the foregoing. It is also constant to its peculiarities; of which the chief are that the stipes or rachis is ramose; the pinnæ scarcely bipinnate, the lowest pinnules only and these only at the base of the frond being separate, the rest irregularly confluent and divided into crowded lobes, of irregular shape, having unequal bristle-tipped serratures. It is a curious dwarf variety.

21. *angustipinnula* (M.). This is a curious form, depauperated to a certain extent and yet hardly affecting its symmetry. The fronds appear to be normally oblong-lanceolate in outline, upwards of a foot in height, with the pinnæ rather distant below; they are distinctly bipinnate. The pinnules stand apart, and are mostly of linear outline, the usual oblong outline being narrowed by the lobes being all and nearly equally depauperately shortened throughout; frequently the basal anterior one is elongated like an auricle, but the rest are nearly uniform, and short. The pinnules are consequently linear (nearly an inch long), with the margins unequally inciso-serrate. In the upper part of the frond, the pinnules become abbreviated into irregular roundish lobes, but with a certain degree of uniformity in size. This plant, if it remains true to these peculiarities, will be a very remarkable variation of the monstrous or depauperated class. In some fronds produced by the same plant the following season to that in which those previously described were formed, the same general character was preserved, only the symmetry was lost, some of the lower pinnules being, like the upper, abbreviated and roundish, instead of linear. It was found by Mr. R. Morris, we believe somewhere in Lancashire. A similar plant to this has been obtained by Mr. Willison of Whitby, Yorkshire, who has sent it under the name of *cystopteroides;* it has the same general character as *angustipinnula,* but the pinnules are scarcely so much narrowed as in the fronds above described. [Plate XLIX B.]

22. *erosa* (Woll.). The peculiarity of this form, which somewhat resembles the *collina* type, resides in the marginal toothing, which is so arranged as to produce a nibbled but scarcely unsymmetrical appearance. The frond is narrowly ovate, bipinnate; the pinnules distant, oblong obtuse, lobed, and the lobes serrated, the serratures being bristle-pointed, and often curved. A peculiar feature of the variety consists in the pinnules being somewhat unequal in size, in the lobes of the pinnules being also unequal, and consequently in the toothing being irregular. It was found by Mr. Wollaston at Tunbridge Wells. Mr. Clowes has found at Windermere a similar plant, but larger and more deeply lobed, the pinnæ being, though very slightly, here and there interrupted.

23. *interrupta* (M.). This, is a variety of the *collina* type, and is of medium size, and of irregular development. The fronds are lanceolate-ovate. The pinnæ, normally composed of close-set oblong obtuse pinnules, are sometimes wanting, sometimes reduced half or more in length, and acquiring an irregularly incised or laciniated margin; sometimes nearly normal towards the tip, and having the pinnules near the base variously reduced, and altered in outline, as well as laciniated. It is as the name implies, one of the interrupted or irregular forms. The fronds are sometimes ramose as well as interrupted. It was found near Harrogate in Yorkshire by Mr. Clapham.

24. *cristata* (M.). This appears to be a dwarfish form, of broad ovate outline. The pinnæ are mostly about twice-forked into short bluntly dilated segments forming a slightly crisped terminal tuft, and the apex of the frond is similarly but rather more deeply divided. When not thus tufted, the apices of the pinnæ are very broad, blunt, and dilated, showing by the branching of the veins a tendency to become forked, so that it is probable the plant may eventually prove to be more thoroughly cristate than our figure represents. It was found near Doncaster by Mr. S. Appleby, and was communicated by Mr. R. Sim of Footscray. [Plate XLIX A.]

25. *glandulosa* (M.). This is a large growing and somewhat erect habited plant, with much the aspect of a large broad *spinulosa*, but differing from that plant in the scales of the stipes being frequently two-coloured and more lanceolate, and in the indusia being fringed with glands. The caudex proves to be decumbent or somewhat creeping.

The stipes varies from about one-third to one-half the entire length of the frond, and is clothed sparingly upwards, more thickly near the base, with ovate bluntish and ovate-lanceolate pointed scales, which are generally of a pale brown, some but not all having a darker central blotch or streak, and many of them, as seen on the growing plant, becoming a good deal appressed to the stipes, whilst a few remain spreading, this peculiarity being of course far less obvious after the fronds are pressed. The stipites, rachides, and under surface of the fronds, are densely covered with stalked glands. The fronds are from two to four feet high, oblong-lanceolate in the larger plants, or ovate-lanceolate in the smaller ones, growing nearly erect around the stout pale-coloured crown which terminates the thick tufted ascending or slowly creeping caudex; bipinnate above, tripinnate below. The pinnæ are ascending, and twisted so as to form nearly a horizontal plane, the lower ones broad and unequally deltoid, the upper lanceolate-ovate, the longest nearly six inches long, and about two inches broad just above the base. The pinnules are lanceolate-ovate, or pyramidately ovate, acute, averaging nearly an inch in length over the greater part of the frond, the posterior ones on the lower pinnæ longest, those of the lowest pinnæ being an inch and three-quarters long; the lower ones stalked, the rest successively decurrent, adnate, and then confluent; they are pinnatifid almost down to the midvein; their lobes oblong, adnate, incised or toothed, the serratures all tipped by a bristle-like point. The fructification is copious over the whole frond, the sori forming two lines on each of the smaller pinnules, or on the lobes of the larger ones, and being covered by indusia, which are fringed with stalked marginal glands. This variety was first discovered by Mr. Bennett, of Brockham, in a boggy part of Ankerbury Hill, near Lydbrook, in the Forest of Dean, Gloucestershire. The same form has since been obtained from —Essex: Epping Forest, *H. Doubleday*. Sussex: wood at Hastings, *J. Stidolph*. Surrey: Barnes (rather less glandular and with more narrow scales), *T. M.* Shropshire: wood below Linley, near Broseley, *G. Maw*. Westmoreland: Windermere, *F. Clowes*. We are indebted to all these gentlemen, as well as to Mr. Purchas, for specimens. The Windermere plant has a more creeping caudex, but it is not otherwise distinguishable from the rest.

We have met with several large-fronded glandular forms of this species, which are barely if at all distinguishable from the present variety. They differ chiefly in having few or none of the broad pallid scales to be found on true *glandulosa,* and also in the less ascending position assumed by the pinnæ. The principal of these are: —1: a smaller frond, with paler scales, and less pyramidate pinnules, but glandular, gathered near Croydon, Surrey, by Mr. J. Hutcheson; it is perhaps referrible to *glandulosa* itself, as we have seen the same much larger under cultivation. 2: a broader large-growing form from Ruthin, Denbighshire, gathered by Mr. T. Pritchard, and a similar one from Castle Malgwyn, Pembrokeshire, gathered by Mr. W. Hutchison, both coming very near *glandulosa.* 3: a glandular form, collected by Dr. Allchin, at Festiniog, which when found was small and like *dumetorum,* but proves to be a large growing form closely resembling *glandulosa.* Other glandular forms of *dilatata* have been gathered at Hampstead Middlesex, at Barnes Surrey, at Chislehurst Kent, and at Windermere; these are somewhat different from the foregoing, and supply connecting links between *glandulosa* and *dilatata.*

THE HAY-SCENTED, or CONCAVE BUCKLER FERN.

LASTREA ÆMULA.

L. fronds triangular or triangular-ovate, spreading tripinnate; pinnules concave; pinnulets pinnatifid, the mucronately serrate lobes curved upwards; scales of the stipes whole-coloured, narrow-lanceolate, fimbriate or laciniate, often contorted; indusium margined with minute sessile glands.

LASTREA ÆMULA, *Brackenridge*, *United St. Explor. Exped.* xvi. 200 (excl. syn. Presl). *J. Smith*, *Cat. Kew Ferns*, 1856. *Moore*, *Handb. Brit. Ferns*, 3 ed. 139. *Johnson*, *Kew Journ. Bot.* ix. 163.

LASTREA FŒNISECII, *Watson*, *Phytol.* ii. 568. *Babington*, *Man. Brit. Bot.* 4 ed. 422. *Deakin*, *Florigr. Brit.* iv. 117, fig. 1614. *Moore*, *Handb. Brit. Ferns*, 2 ed. 132; *Id.*, *Ferns of Gt. Brit. Nature Printed*, t. 27. *Sowerby*, *Ferns of Gt. Brit.* 27, t. 14.

LASTREA RECURVA, *Newman*, *Nat. Alm.* 1844, 23; *Id.*, *Hist. Brit. Ferns*, 2 ed. 225.

LASTREA CONCAVA, *Newman MS.* (Hist. Brit. Ferns, 2 ed. 235).

NEPHRODIUM FŒNISECII, *Lowe* (*R. T.*), *Cambr. Phil. Trans*, iv. 7.

ASPIDIUM ÆMULUM, *Swartz*, *Schrad. Journ. Bot.* 1800, ii. 42; *Id.*, *Syn. Fil.* 60, 257. *Willdenow*, *Sp. Plant.* 283. *Sprengel*, *Syst. Veg.* 108. *Lowe*, *Nat. Hist. Ferns*, vii. t. 17.

ASPIDIUM FŒNISECII, *Kunze*, *Lin.* xxiii. 226. *Fée*, *Gen. Fil.* 292.

ASPIDIUM ODORATUM, *Lowe MS.*: *Hb. Hooker.*

ASPIDIUM DILATATUM, *v.* CONCAVUM, *Babington.*

ASPIDIUM SPINULOSUM, *Bentham*, *Handb. Brit. Fl.* 630, in part.

POLYPODIUM ÆMULUM, *Aiton*, *Hort. Kew*, iii. 466; according to an authentic specimen in Banksian Herbarium.

ALLANTODIA ÆMULA, *Desvaux*, *Ann. de Soc. Linn. Paris*, vi. 265.

ASPIDIUM RECURVUM, *Bree*, *Phytol.* i. 773.

ASPIDIUM DILATATUM, *v.* RECURVUM, *Bree*, *Mag. Nat. Hist.* iv. 162.

ASPIDIUM SPINULOSUM γ. *Hooker & Arnott*, *Brit. Fl.* 7 ed, 586.

LOPHODIUM FŒNISECII, *Newman*, *Phytol.* 1851, *App.* xvi.; *Id.*, *Brit. Ferns*, 3 ed. 135 (excl. syn. Asp. dumetorum).

LOPHODIUM RECURVUM, *Newman*, *Phytol.* iv. 371.

Caudex stout, densely scaly, tufted, erect or sometimes decumbent, formed of the bases of the fronds surrounding a woody axis. *Scales* narrow-lanceolate, pale ferruginous, of one colour, variously and sparingly fimbriate or lacerate on the margin, and generally contorted. *Fibres* long, stout, wiry, branched, dark brown, tomentose.

Vernation circinate.

Stipes usually about half the entire length of the frond, rigid, moderately stout, brownish-purple from the base upwards, furnished plentifully with subulately-lanceolate entire fimbriate or lacerate scales of a pale rusty-brown colour; terminal, and adherent to the caudex. *Rachis* greenish, furnished with fewer and smaller scales, and as well as the stipites and secondary rachides bearing numerous small sessile spherical glands.

Fronds numerous, from one to two feet high including the stipes, and from about five to eight inches across, sometimes smaller, of a rich bright green, somewhat paler beneath, drooping, the upper surface crispy; triangular, or elongate-triangular, or sometimes ovate, tripinnate, the lower surface sprinkled with minute sessile glands. *Pinnæ* opposite or sub-opposite, the lowest usually, but not always the longest, broadly and unequally deltoid, the pinnules on the posterior side being larger than those on the anterior; the succeeding pinnæ become gradually narrower and less oblique. *Pinnules* pyramidately-triangular, or obliquely-oblong, the basal posterior ones of the lowest pinnæ much longer than the rest, and divided into ovate-oblong or oblong pinnulets, the largest of which are deeply pinnatifid, the lobes being oblong serrated. The basal pinnæ, pinnules, and pinnulets are all stalked, the upper ones becoming in gradation sessile and then decurrent. The margins of the pinnules and lobes are mucronately toothed, and these margins are turned upwards from the plane of the spreading or drooping frond, so that all the ultimate divisions are concave, and the entire frond has a beautiful crispy appearance, which, together with its lively colour and graceful habit, render it one of the most ornamental of the robust Ferns.

Venation of the pinnulets consisting of a dark-coloured flexuous *vein* formed of a branch from the *costa* or midvein of the primary pinnule; this produces short lateral forked *venules*, the anterior branch of which bears a sorus below its apex; all the *veinlets* terminate within the margin.

Fructification on the back of the frond occupying the whole under surface. *Sori* round, numerous, indusiate, forming two rows along each of the pinnules and pinnulets, placed near to the midvein, often becoming confluent. *Indusium* reniform, its margin jagged and uneven, and sparingly furnished with sessile glands. *Spore-cases* numerous, brown, obovate. *Spores* oblong, granulated.

Duration. The caudex is perennial. The fronds of one season endure until after the earlier ones of the following year are produced, so that the plant is evergreen. The growth takes place in succession from the month of May onwards till autumn.

This beautiful plant is quite distinct both in character and aspect from *Lastrea dilatata* with which it is sometimes associated. Its fragrance, which is a remarkable characteristic, is powerful, resembling that of new hay, becoming too, like that, developed by the desiccation of the plant, and retaining its strength for a length of time in the dried specimens of the herbarium. The scales of the stipites differ from those of *Lastrea dilatata,* both in size, form, and number, being in *Lastrea æmula* fewer, narrower, and for the most part either fringed or lacerate at the margin, pale brown, and concolorous. The fronds too are much more decidedly evergreen than in *L. dilatata;* and, as already mentioned, they have this peculiarity, that they decay from above downwards, and not from the base of the stipes upwards as *dilatata* does. The indusium, moreover, is bordered with sessile, not with stalked glands. In ordinary cases, the triangular outline and concave crispy surface of the fronds will suffice to distinguish the plant, without recourse to the more minute characters residing in the scales and indusia. The plant varies with more elongated fronds, approaching to ovate in outline; while it is to be remembered some forms of *dilatata* are decidedly triangular in outline, so that the form of the frond alone must not be implicitly relied on.

This Fern is found plentifully in the Peninsula formed by the counties of Cornwall, Devon, and Somerset; and it also occurs in several parts of Sussex. It has not been seen in the south-eastern portion of England, but has been gathered in several parts of Wales, in the Isle of Man, in North Lancashire, in Cumberland and Northumberland, and both on the eastern and western sides of Yorkshire. In

Scotland it occurs on the shores of Loch Lomond, on the west coast of Argyleshire, in the Isles of Arran and Mull, in Orkney and the outer Hebrides. In Ireland it is abundant; and it is found in the Channel Isles. The range in altitude appears to extend from the coast level, to an elevation of about 600 feet. It prefers shady rocky habitats. The following stations are recorded:—

Peninsula.—Cornwall: Penzance; Penryn; St. Michael's Mount; Helston; Lostwithiel; Redruth; Truro, and throughout the county. Devonshire: Chambercombe; Ilfracombe; Linton; Hartland, *Rev. J. M. Chanter;* Parracombe Hill, *R. J. Gray;* Barnstaple; Marwood and Eastdown, *F. Mules;* Clovelly; Helsworthy, *Rev. W. S. Hore;* Devil's Tor, Dartmouth; Bickleigh Vale; Shaugh Vale, *R. J. Gray,* etc. Somersetshire: Selworthy.

Channel.—Sussex: Tunbridge Wells; Ardingly; Balcombe; West Hoathly.

Severn.—Herefordshire. Shropshire: Coalbrookdale, *G. Maw.*

Humber.—Yorkshire: Settle, *J. Tatham;* Scarborough, *W. Bean.*

Tyne.—Northumberland: Embleton; Dirrington Law, *Dr. Johnston.*

Lakes.—Cumberland: St. Bee's Head. North Lancashire: Conistone. Isle of Man, *Dr. Allchin.*

S. Wales.—Glamorganshire: Melincourt waterfall, *E. Young.* Pembrokeshire: Castle Malgwyn, Llechryd, *W. Hutchison.*

N. Wales.—Anglesea: Holyhead, *G. Maw.* Merionethshire. Carnarvonshire: Snowdon district, *Dr. Allchin.*

E. Highlands.—Forfarshire: Baldovan, Kinnordy, *W. Gardiner.*

W. Highlands.—Dumbartonshire: Banks of Loch Lomond. Argyleshire: Glen Gilp; Ardrishiag, *T. M.;* Campbelton, *A. Tait;* Loch Swin, on the west coast, *Mrs. Shaw.* Inverness-shire: The Craig, south side of Loch Moidart, *Rev. T. F. Ravenshaw.* Arran: wooded rocks between Brodick and Corrie, and between Lamlash and Whiting Bay, *Dr. Balfour.* Isle of Mull: Tobermory, *W. Tanner.*

N. Isles.—Orkney: Hoy, rather common, *T. Anderson;* Walkmill Bay, *J. T. Syme.*

W. Isles.—N. Uist, *Dr. Balfour.*

Ulster.—Antrim: Fairhead; near Cushendall. Londonderry: near Coleraine; Rushbrook near Londonderry; Garvagh. Donegal: Banks of Lough Swilly; Milroy Bay; Arrigal Hill, near Donegal; Killybegs, *R. Barrington;* near Lough Eske, *R. B.*; about Lough Derg.

Connaught.—Sligo: between Sligo and Manorhamilton, *E. Newman;* Lough Gill, *R. Barrington.* Mayo: Foot of Nephin; Coraan, Achill; Newport; Westport, etc., *E. Newman.* Galway: about Clifden; about Roundstone and Ballynahinch; near Oughterard; Connemara.

Leinster.—Wicklow: Seven Churches, abundant, *D. Moore;* Glendalough, abundant and luxuriant; Powerscourt, *J. Ball;* Great Sugar-loaf, *J. R. Kinahan.*

Munster.—Waterford: Ballyquin, plentiful; Clonmel, *J. Sibbald;* Ardmore, on sea cliffs sparingly; Foxe's Cove, etc., *J. R. Kinahan.* Clare: Lough Graney; and near Feacle, *J. R. Kinahan;* near Loop-head. Cork: Glanmire, near Cork, *J. Carroll;* woods about Glengarriff. Kerry: on the mountains and in the woods of Kerry, especially about Killarney, Dinis Island, Cromaglaun, and O'Sullivan's Cascade; Kenmare.

Channel Isles.—Guernsey, *J. James.*

Beyond the limits of the United Kingdom, the species is found, we believe, only in the Atlantic Isles off the African coast, namely the Azores, Madeira, and the Cape de Verd Isles.

The Hay-scented Buckler Fern grows freely in cultivation, when planted in a porous soil of loam, peat, and sand; and is to be considered as one of the most ornamental of our native species. Its decidedly evergreen character and moderate size, no less than its elegant crispy appearance, adapt it especially for greenhouse culture, or for a Wardian case of tolerable size: and in such situations it requires no unusual or special treatment. It is in fact, a perfectly manageable species. The plants may be multiplied by separating the crowns.

It does not yield much variety. We have received from Mr.

Moore of Glasnevin, a form found near Cushendall in Antrim, in which the secondary pinnules are more confluent than usual, the lobes irregularly shortened (*angustipinnula*), and the teeth more evidently aristate. The following variety has also been observed :—

1. *interrupta* (Claph.). The fronds are variously depauperated, the pinnæ or more frequently the pinnules, or sometimes both, being interruptedly abbreviated and misshapen, producing considerable irregularity in the details of the frond, but not much disturbance of its general outline. It was found by Mr. Clapham at Hackness near Scarborough, Yorkshire.

END OF VOL. I.

BRADBURY AND EVANS, PRINTERS, WHITEFRIARS

Plate I
Polypodium vulgare
d by Henry Bradbury

Plate II.

POLYPODIUM VULGARE VARS:

A. ACUTUM. B. AURITUM.

by Henry Bradbury.

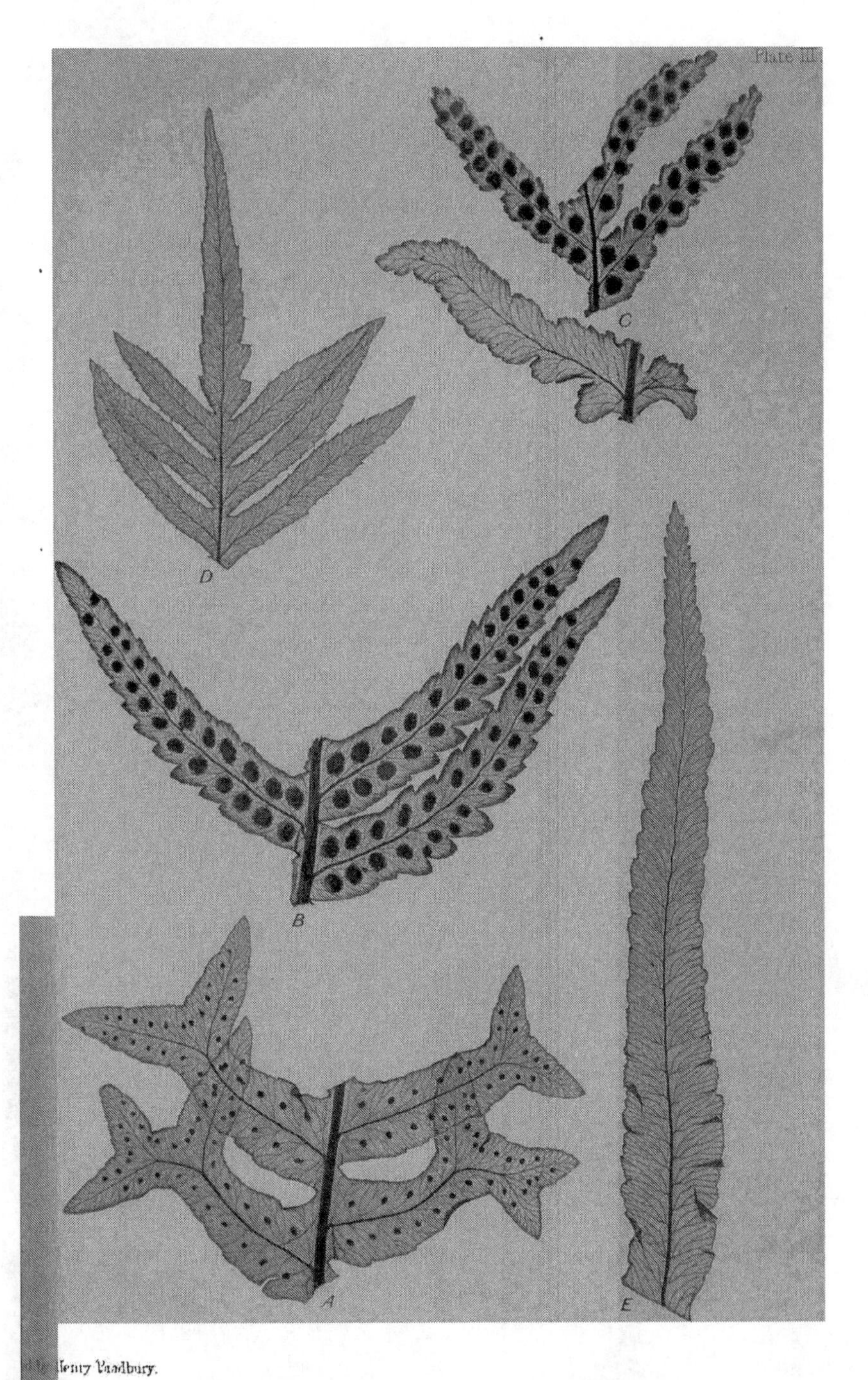

Henry Bradbury.

POLYPODIUM VULGARE, VARS: A. BIFIDUM,
B. SERRATUM, C. SINUATUM, D. DENTICULATUM, E. CRENATUM.

Plate IV

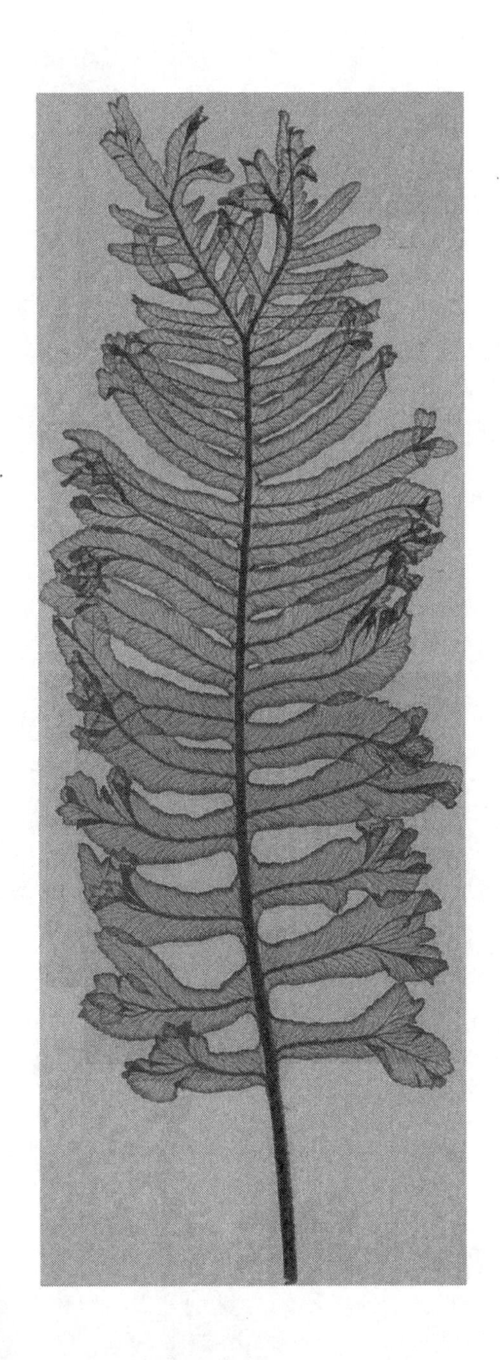

POLYPODIUM VULGARE, VAR: CRISTATUM.

by Henry Bradbury

Plate V.

POLYPODIUM VULGARE, VAR. SEMILACERUM.

[...]d by Henry Bradbury.

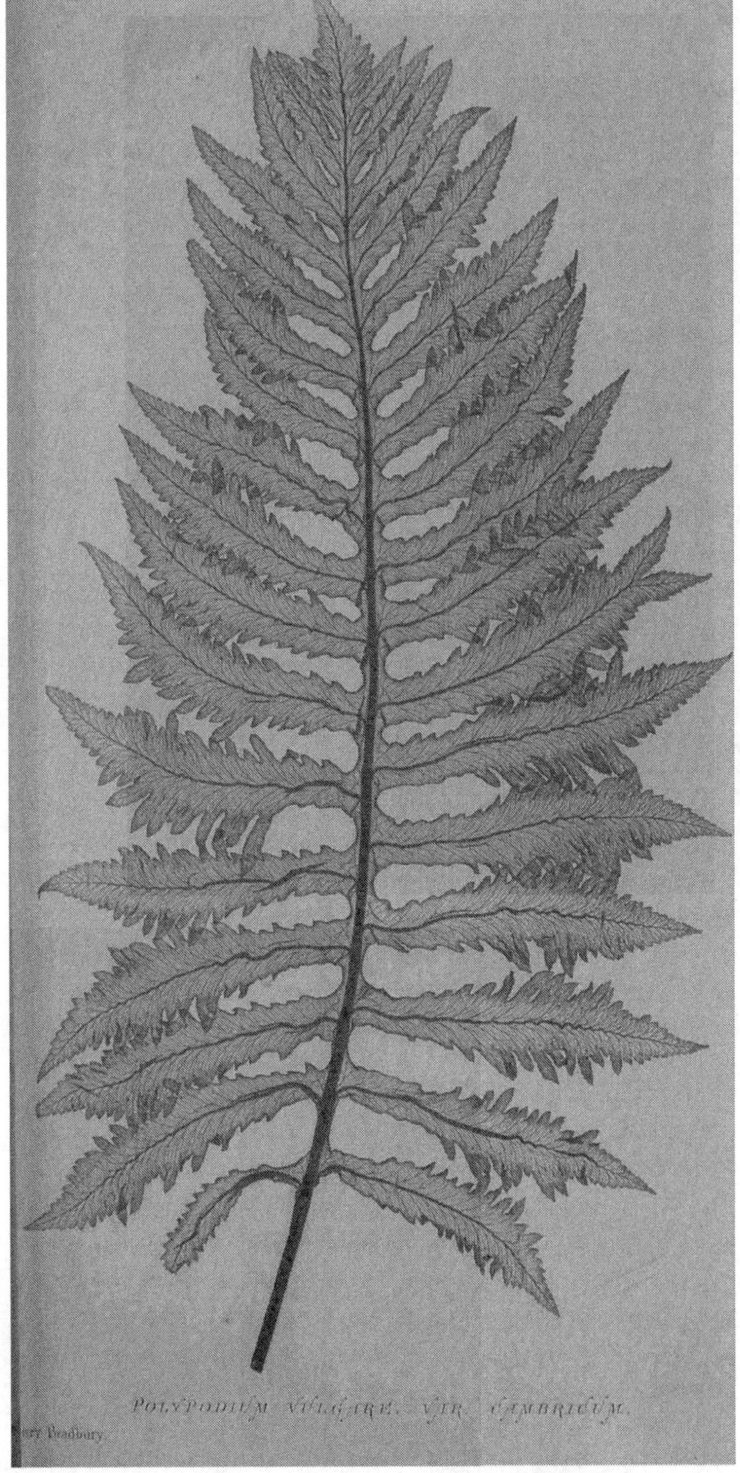

POLYPODIUM VULGARE. VAR. CAMBRICUM.

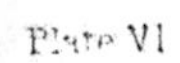

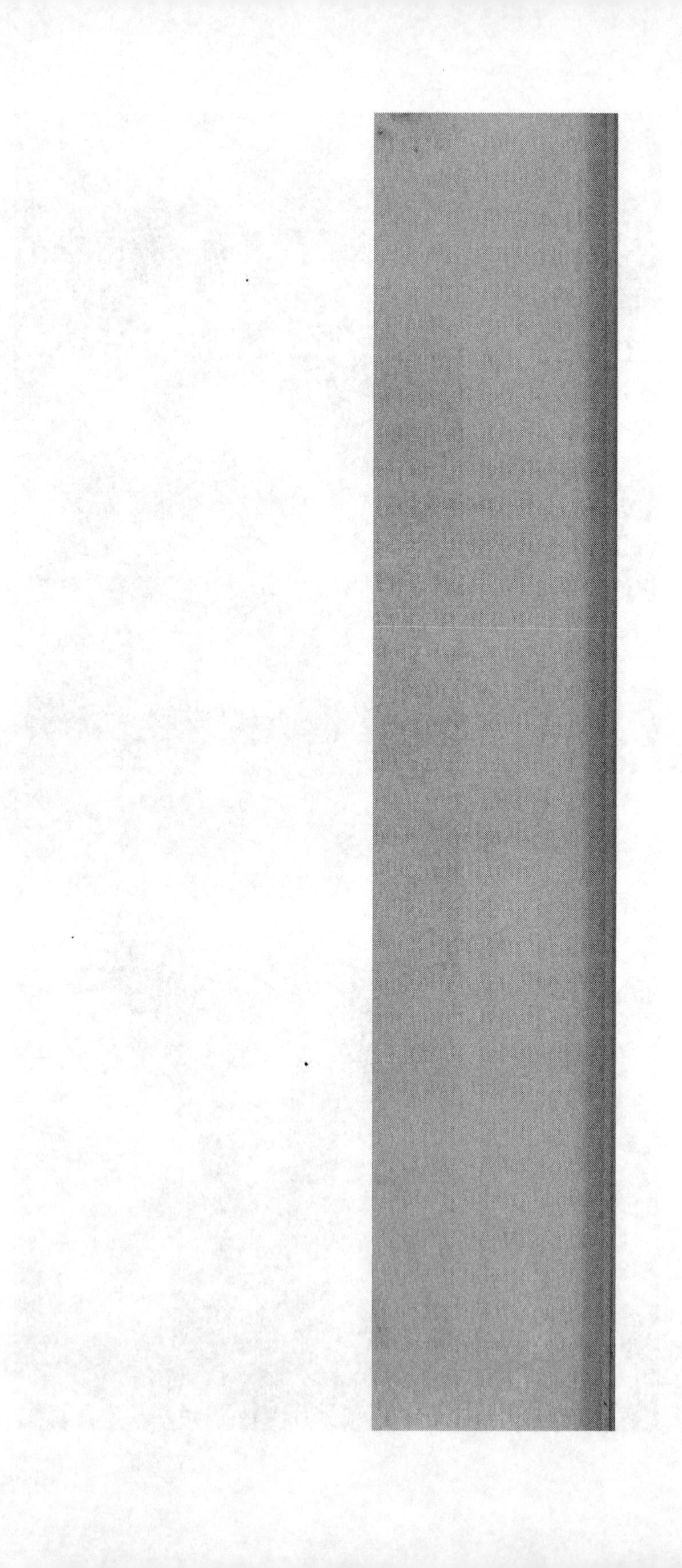

Plate VII

POLYPODIUM VULGARE, VAR: OMNILACERUM.

by Henry Bradbury

Plate VIII.

POLYPODIUM PHEGOPTERIS.

Plate IX

"ly Henry Bradbury.

POLYPODIUM ALPESTRE. VAR. FLEXILE.

Plate XI
Polypodium Dryopteris
by Henry Bradbury

Plate XII.

POLYPODIUM ROBERTIANUM.

Plate XIII

Printed by Henry Bradbury

ALLOSORUS CRISPUS.

Plate XIV.

Plate XV.

Plate XVI.
POLYSTICHUM ACULEATUM.
Henry Bradbury

Plate XVII.

POLYSTICHUM ACULEATUM, VARS. A. LOBATUM, B. ARGUTUM.

Henry Bradbury

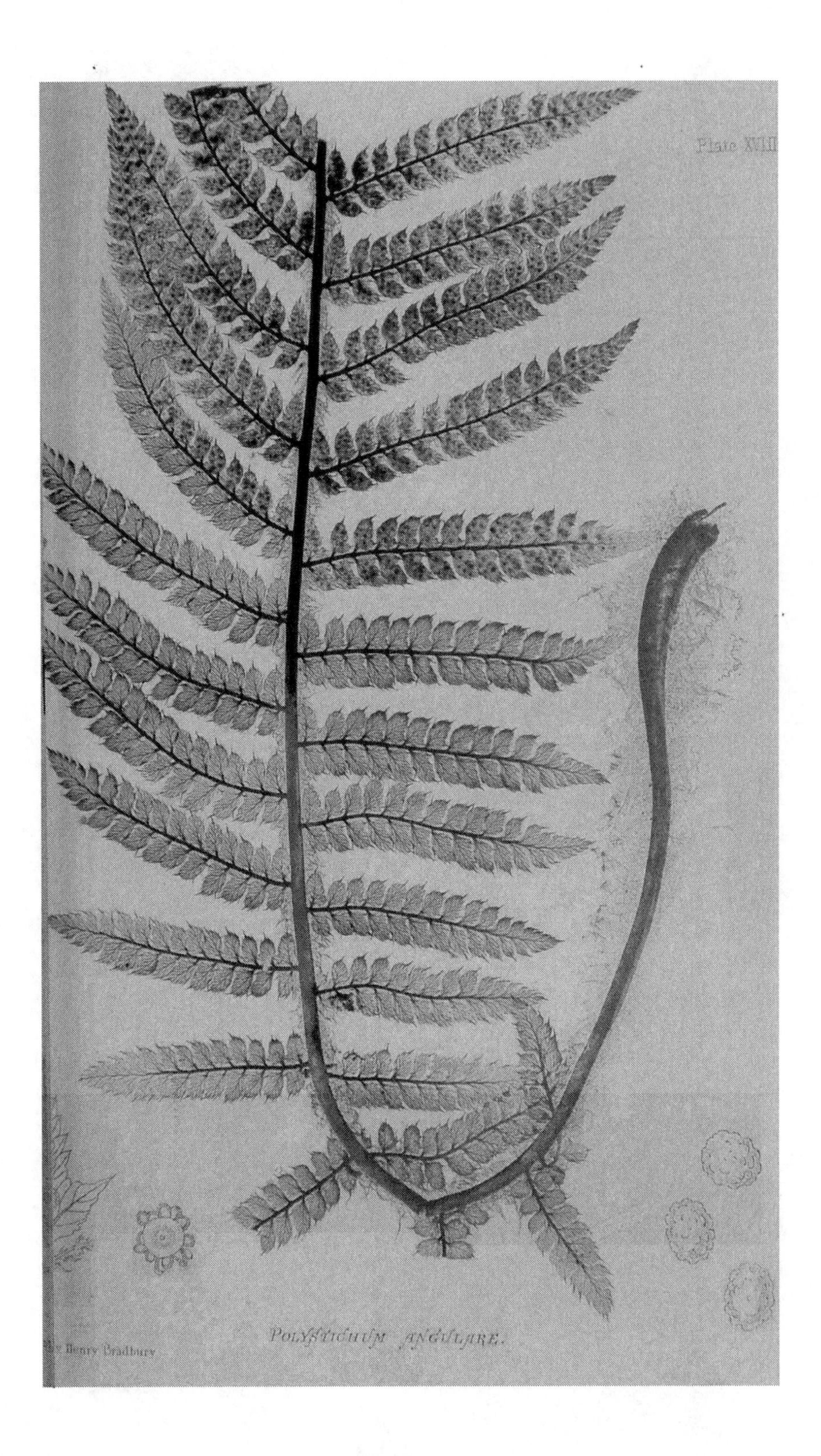

POLYSTICHUM ANGULARE.

Plate XIV.

POLYSTICHUM ANGULARE, VARS:
A. SUBTRIPINNATUM, B. INTERMEDIUM, C. BISERRATUM,
D. HASTULATUM, E. TRIPINNATUM.

Henry Bradbury

Plate XX

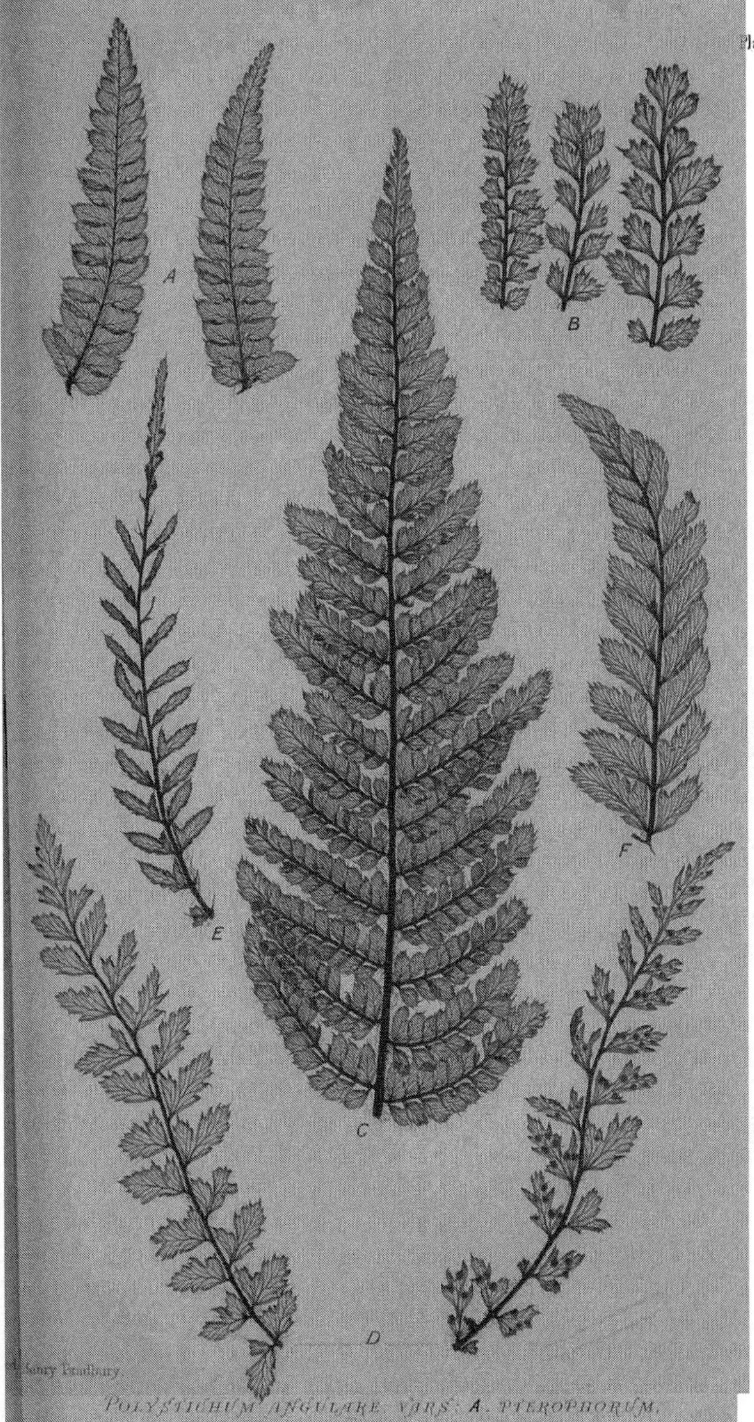

POLYSTICHUM ANGULARE. VARS: A. PTEROPHORUM. B. GRANDIDENS. C. DENSUM. D. IRREGULARE. E. LINEARE. F. PLUMOSUM.

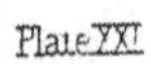
Plate XXI

POLYSTICHUM ANGULARE VAR. IMBRICATUM.

Plate XXII.

by Hy Bradbury. POLYSTICHUM ANGULARE, VAR. ALATUM.

Plate XXIII.

POLYSTICHUM ANGULARE, VAR.
PROLIFERUM WOLLASTONI.

...d by Henry Bradbury

Plate XXIV

POLYSTICHUM ANGULARE VAR. DISSIMILE.

... by Henry Bradbury

Plate XXV.

POLYSTICHUM ANGULARE, VARS:
A. GRANDIDENS, B. GRACILE.

Henry Bradbury.

Plate XXVI.

POLYSTICHUM ANGULARE VAR. CONFLUENS.

by Henry Bradbury

Plate XXVII.

Polystichum angulare, vars:

A. Cristatum B. Polydactylum.

ted by Henry Bradbury

Plate XXVIII.

POLYSTICHUM ANGULARE, VAR KITSONIAE.

Nature Printed by Henry Bradbury.

by Henry Bradbury

LASTREA THELYPTERIS.

LASTREA MONTANA.

...ted by Henry Bradbury

Printed by Henry Bradbury.

LASTREA MONTANA, VAR. TRUNCATA.

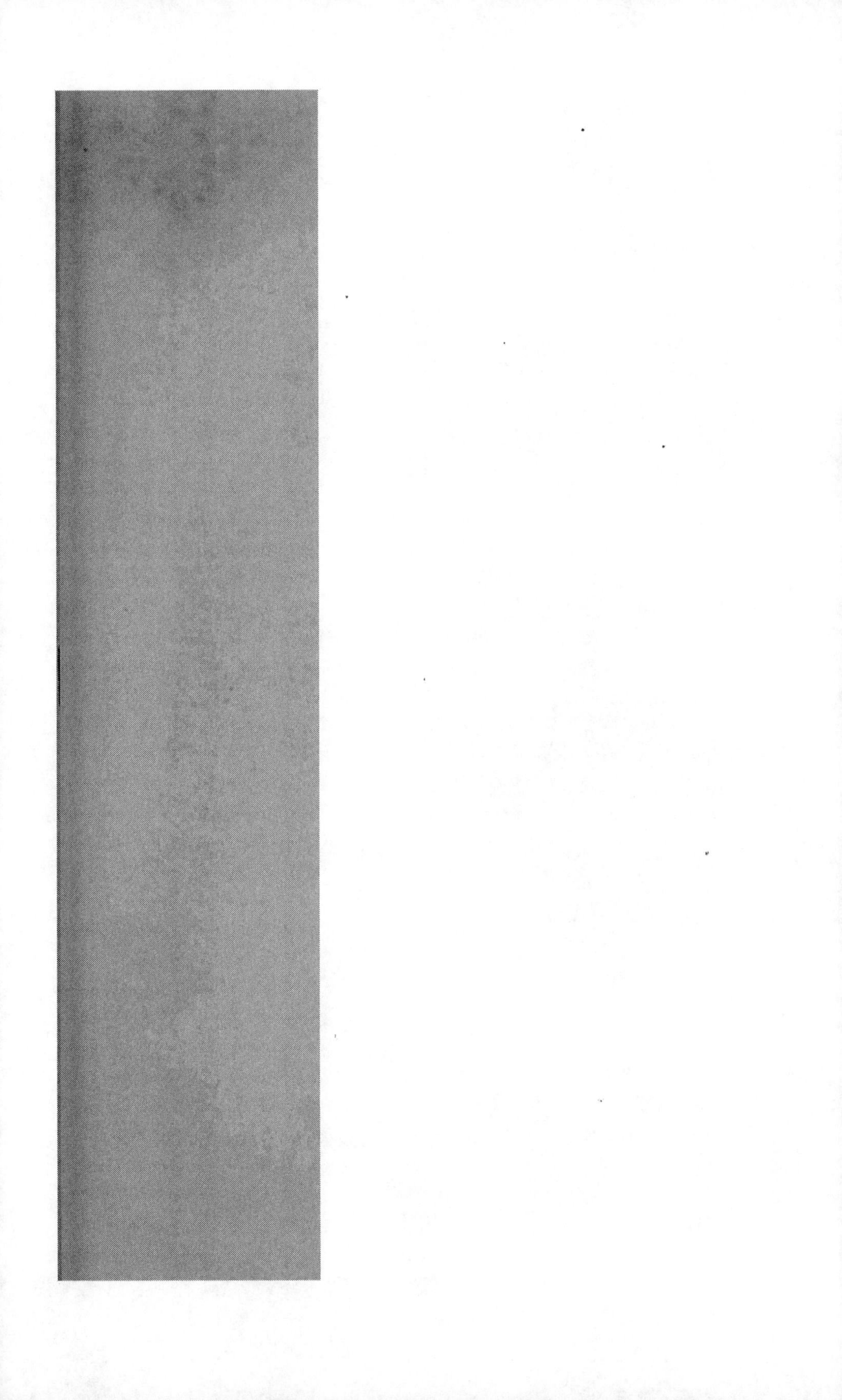

Plate XXXII
LASTREA FILIX-MAS.
Printed by Henry Bradbury.

LASTREA FILIX-MAS, VARS.

A. PALEACEA. B. INCISA. C. PALEACEO-LOBATA.

Plate XXXIV

LASTREA FILIX-MAS, VARS: A. ELONGATA, B. PRODUCTA.

Printed by Henry Bradbury

Plate XXXV.

Lastrea Filix-mas, var. pumila.

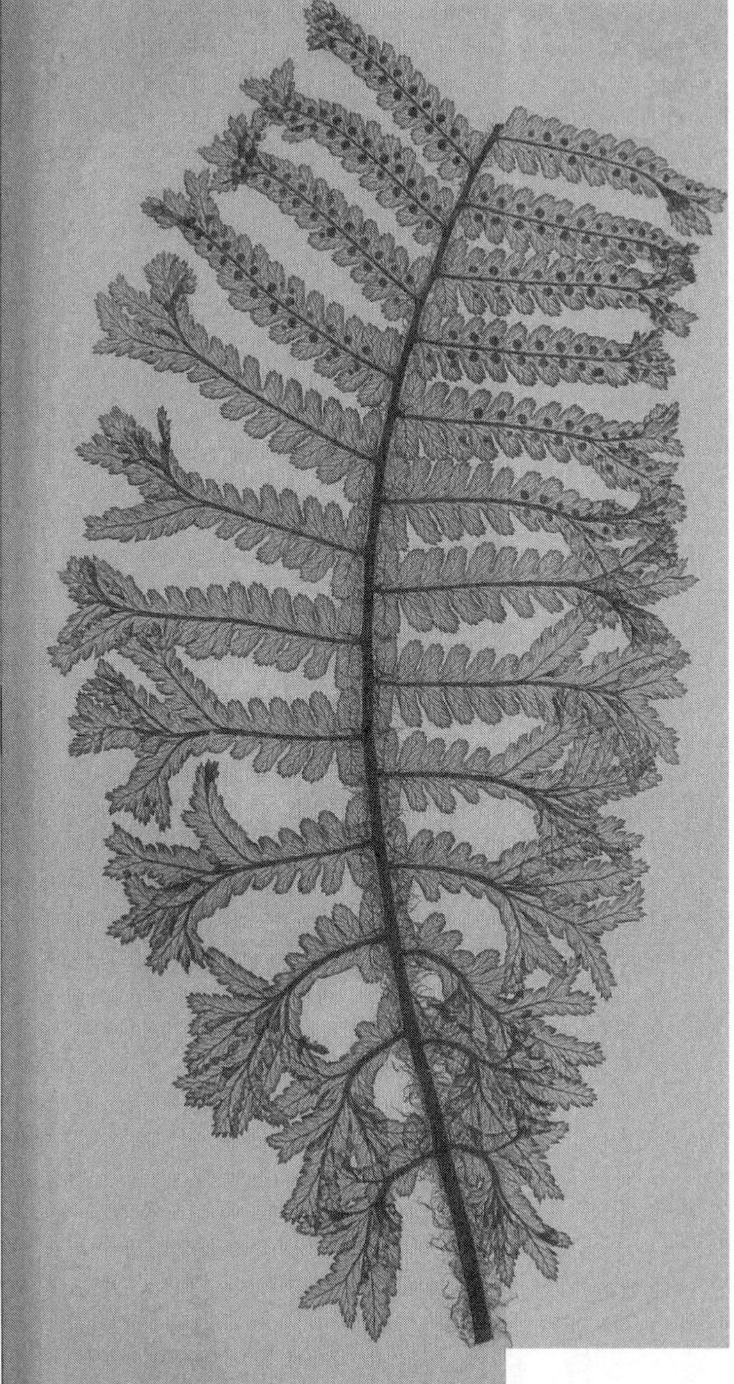
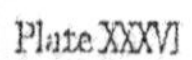

Plate XXXVI

LASTREA FILIX-MAS, VAR: CRISTATA.

Printed by Henry Bradbury

Plate XXXVII.

LASTREA FILIX MAS, VAR: POLYDACTYLA.

Printed by Henry Bradbury.

Plate XXXVIII.

LASTREA FILIX MAS, VAR. SCHOFIELDII.

ure Printed by Henry Bradbury.

LASTREA RIGIDA.

[...]nted by Henry Bradbury

LASTREA CRISTATA.

Plate XLI.

LASTREA CRISTATA VAR. ULIGINOSA.

Printed by Henry Bradbury.

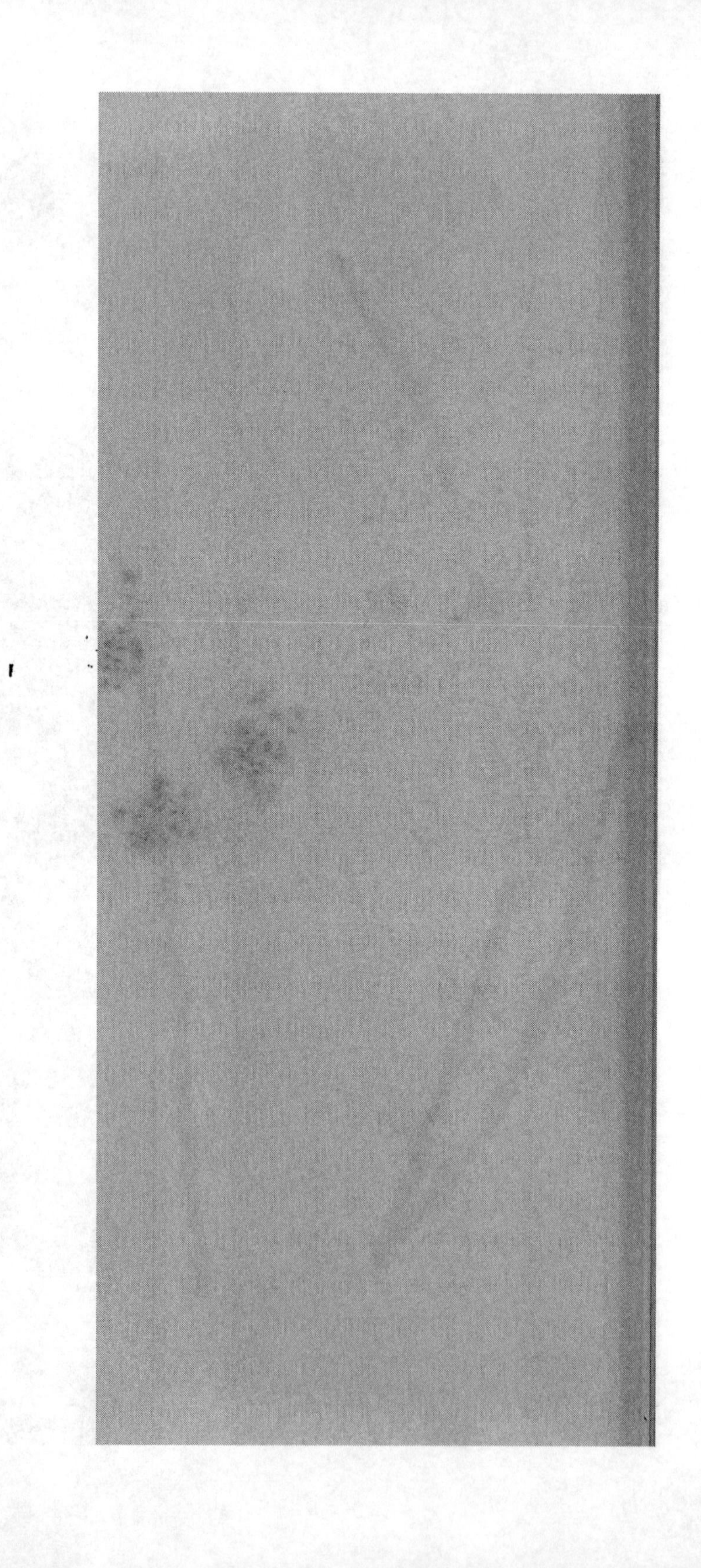

Plate XLII.

Printed by Henry Bradbury.

LASTREA CRISTATA, VAR. SPINULOSA.

Plate XLIII

[Na]ture Printed by Henry Bradbury.

LASTREA DILATATA.

LASTREA DILATATA, VAR. GLANDULOSA.

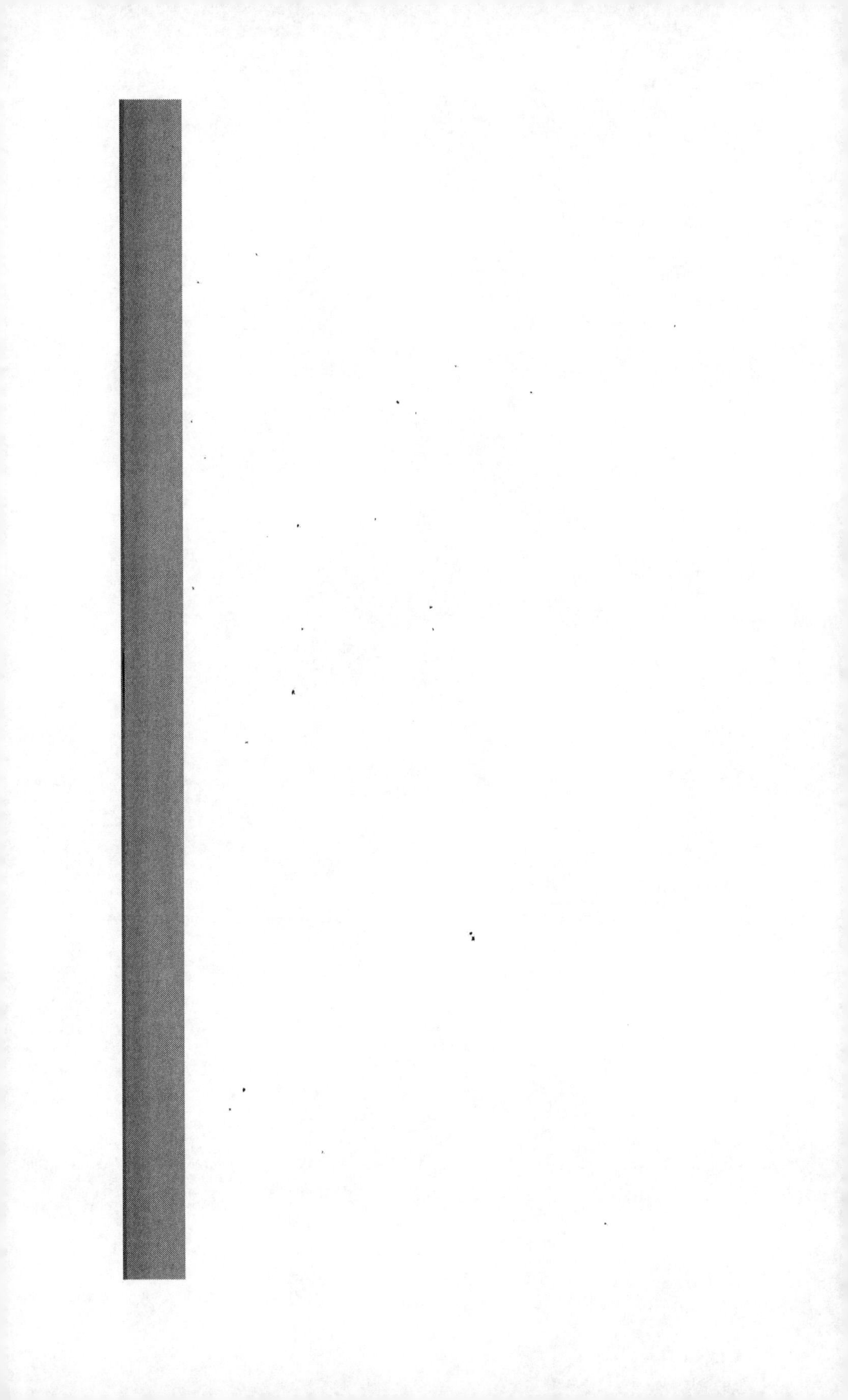

Plate XLV

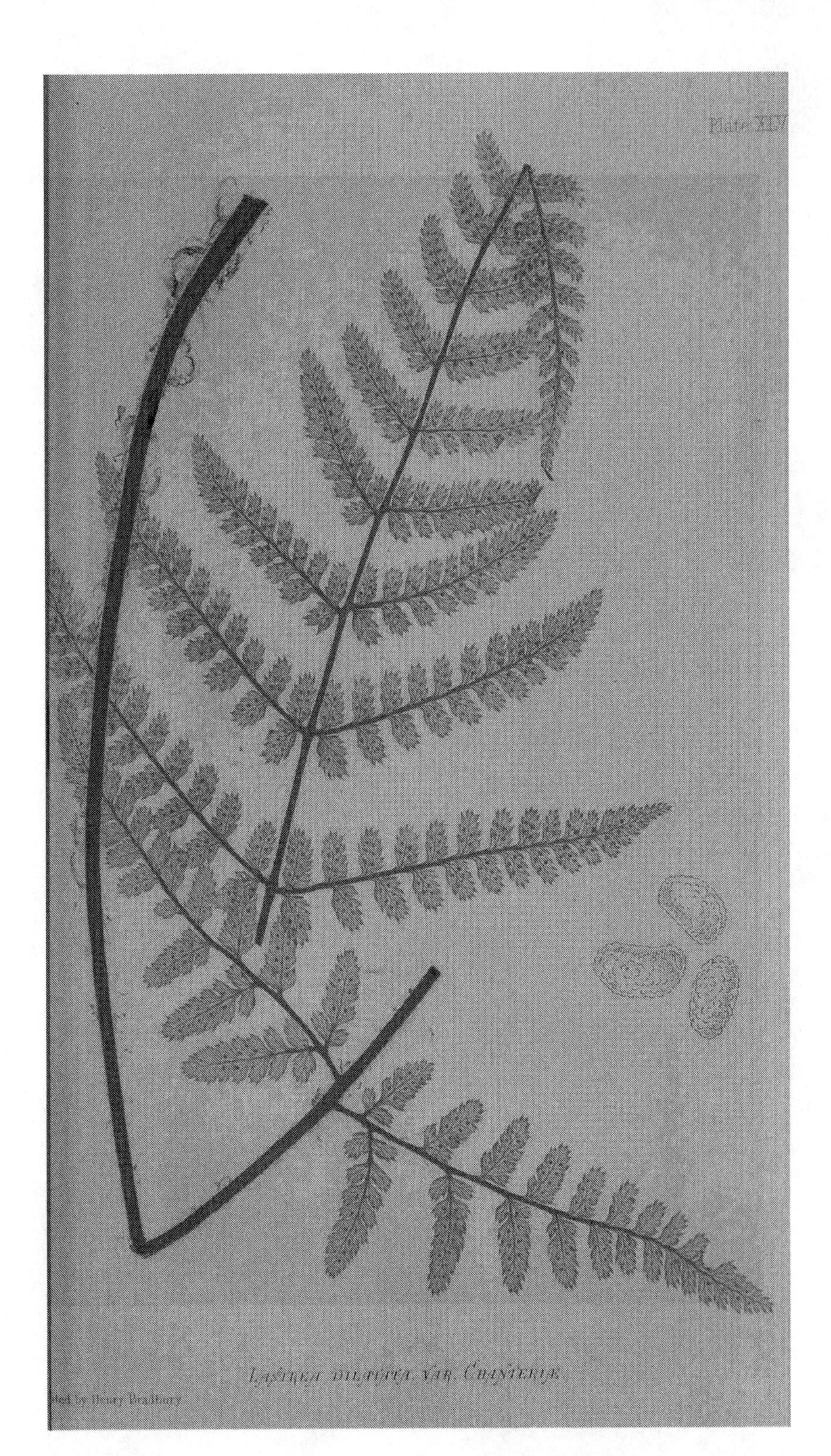

LASTREA DILATATA, VAR. CHANTERIÆ.

ted by Henry Bradbury

Plate XLVI.

Lastrea Dilatata. Var. Nana.

[...]ted by Henry Bradbury.

Plate XLVII.

LASTREA DILATATA, VAR, COLLINA.

Plate XLVIII.

LASTREA DILATATA, VAR: DUMETORUM.

Nature Printed by Henry Bradbury.

Plate XLIX.

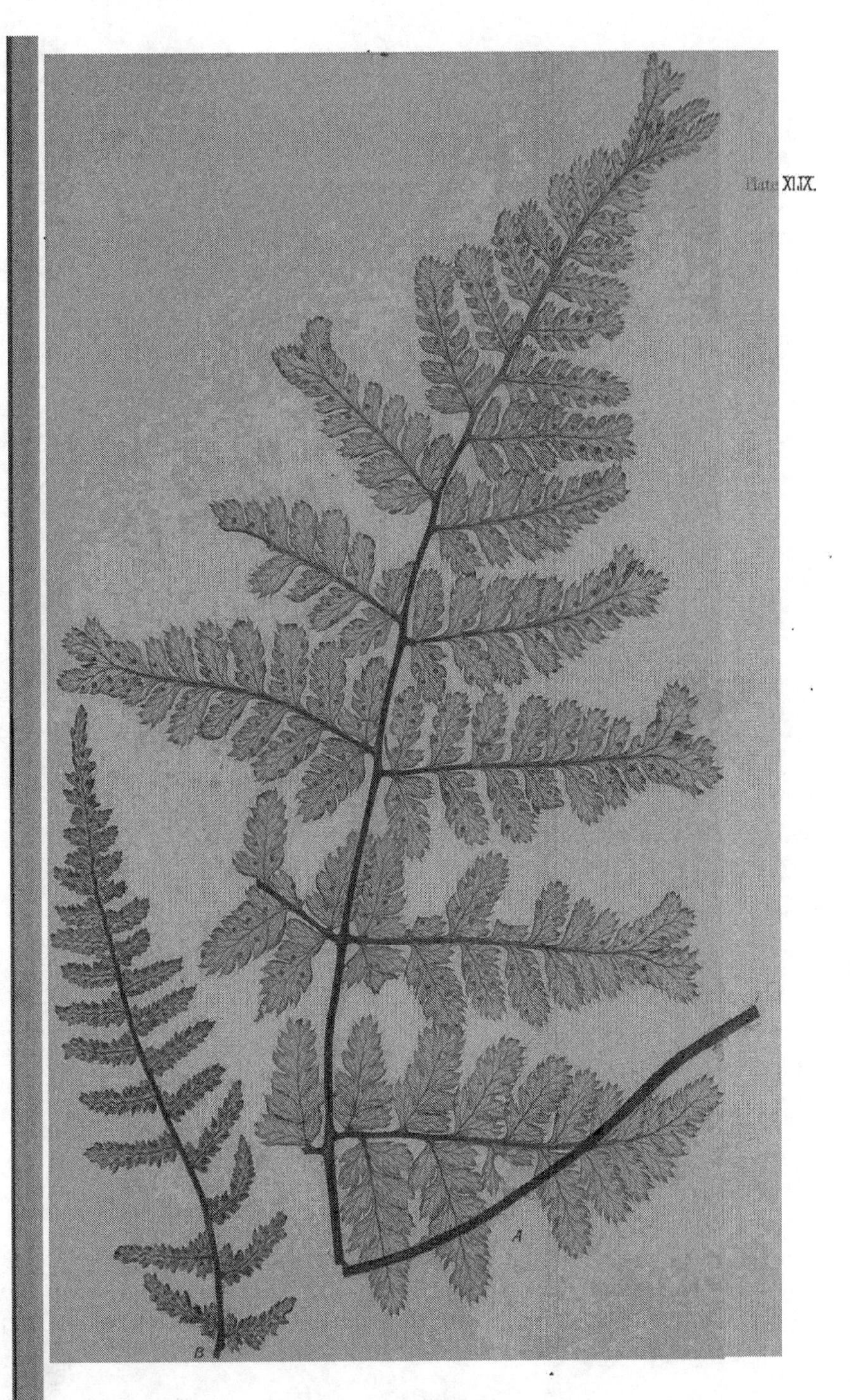

LASTREA DILATATA, VARS: A. CRISTATA. B. ANGUSTIPINNULA.

...ted by Henry Bradbury

Plate L.

...nted by Henry Bradbury.

LASTREA DILATATA, VAR. LEPIDOTA.

LASTREA ÆMULA.

**** A General Index to the two volumes will be given at the end of the Work.*

Volume II. will be published in December.

CPSIA information can be obtained
at www.ICGtesting.com
Printed in the USA
LVHW082209140323
741644LV00011B/542